"十四五"时期水利类专业重点建设教材

港口航道与海岸工程专业英语

主 编 姚宇 李江夏 任韧希子

·北京·

内 容 提 要

本书为高等院校港口航道与海岸工程专业的专业课教材，全面介绍了港口航道与海岸工程中相关知识点的专业英语表达。全书共分为 4 个章节，每章包含 4～5 篇课文。第 1 章港口工程，介绍了港口规划与布置、码头功能分类、码头结构分类、荷载分类、货物装卸设备的相关知识；第 2 章航道工程，介绍了河道类型、河床演变、疏浚工程、整治建筑物、船闸的相关知识；第 3 章海岸工程，介绍了海浪、防波堤、海堤和丁坝、海岸地貌、海岸输沙和海滩养护的相关知识；第 4 章海洋工程，介绍了海上平台、波浪能、海底管线、海洋灾害的相关知识。每篇课文配有图片说明、专业词汇列表和全文翻译。

本书可作为高等院校港口航道与海岸工程相关专业的本科生和研究生教材，也可供相关专业的教师、工程技术人员和科研人员参考。

图书在版编目（CIP）数据

港口航道与海岸工程专业英语 / 姚宇，李江夏，任韧希子主编. -- 北京：中国水利水电出版社，2024. 10. -- （"十四五"时期水利类专业重点建设教材）. ISBN 978-7-5226-2887-5

Ⅰ. U6；P753

中国国家版本馆CIP数据核字第2024U1N989号

书　名	"十四五"时期水利类专业重点建设教材 **港口航道与海岸工程专业英语** GANGKOU HANGDAO YU HAI'AN GONGCHENG ZHUANYE YINGYU
作　者	主编　姚　宇　李江夏　任韧希子
出版发行	中国水利水电出版社 （北京市海淀区玉渊潭南路1号D座　100038） 网址：www.waterpub.com.cn E - mail：sales@mwr.gov.cn 电话：（010）68545888（营销中心）
经　售	北京科水图书销售有限公司 电话：（010）68545874、63202643 全国各地新华书店和相关出版物销售网点
排　版	中国水利水电出版社微机排版中心
印　刷	天津嘉恒印务有限公司
规　格	184mm×260mm　16开本　6.75印张　214千字
版　次	2024年10月第1版　2024年10月第1次印刷
印　数	0001—2000 册
定　价	**28.00元**

凡购买我社图书，如有缺页、倒页、脱页的，本社营销中心负责调换

版权所有·侵权必究

前言

随着我国"海洋强国"建设和"一带一路"倡议的推进，大批从事港口航道与海岸工程建设的企业加快"走出去"步伐，不断开拓海外市场和增强国际化经营能力，对具备良好的英语沟通能力的专业人才的需求也不断增加。近几年，我国在港口航道与海岸工程建设中从发达国家引进了大批先进技术和软硬件设备，消化新技术和维护使用好新设备需要毕业生具备相应的阅读和翻译技能。

在新形势下，大学英语的教学如果仅仅停留在通用英语层面，已不能适应国家和行业的发展需求。开展语言与专业相结合的专业英语教学，发挥语言的工具性作用，是当今复合型人才培养的必然要求，也是衔接通用英语教学和专业课教学不可或缺的教学环节。因此，为了培养具备国际视野，能够在跨文化背景下进行专业交流的港口航道与海岸工程毕业生，亟需一本与时俱进的专业英语教材供本专业学生学习。

本书立足于港口航道与海岸工程中的基本理论、工程技术以及施工工艺，精选了四个主要领域的内容，包括港口工程、航道工程、海岸工程和海洋工程。通过采用专业英语中的常用词、词组、句型和文体，系统地介绍了相关领域的主要专业知识点。同时，本书最大的特点是大量采用绘图的形式，图文并茂地展示专业词汇的实际运用，以加强读者的理解和记忆。书中介绍的内容不仅强调与实际应用的紧密结合，同时又兼顾展示专业领域内最前沿的技术和发展动态。本书既可用于高等院校港口航道与海岸工程相关专业的本科生和研究生教材，也可供相关专业的教师、工程技术人员和科研人员参考。

本书由长沙理工大学主持编写，各章编写分工如下：第1章由李江夏撰写，第2章由任韧希子撰写，第3、4章由姚宇撰写；全书由姚宇统稿和修订。

由于编者水平有限，书中难免存在疏漏和不足之处，敬请广大读者批评指正。

编 者
2024年6月

Contents

Preface

Chapter 1 Port Engineering 1

  Lesson 1.1 Port planning and layout 1

  Lesson 1.2 Classification of terminals by function 5

  Lesson 1.3 Classification of berth structures 8

  Lesson 1.4 Types of loads 13

  Lesson 1.5 Cargo handling equipment 17

Chapter 2 Waterway Engineering 21

  Lesson 2.1 Stream patterns 21

  Lesson 2.2 Fluvial processes 25

  Lesson 2.3 Dredging engineering 28

  Lesson 2.4 Regulating structures 32

  Lesson 2.5 Locks 36

Chapter 3 Coastal Engineering 40

  Lesson 3.1 Ocean waves 40

  Lesson 3.2 Breakwater, seawalls and groynes 44

  Lesson 3.3 Coastal morphology 49

  Lesson 3.4 Coastal sediment transport and beach nourishment 52

Chapter 4 Ocean Engineering 56

  Lesson 4.1 Offshore platforms 56

  Lesson 4.2 Wave energy 61

  Lesson 4.3 Submarine pipelines 65

  Lesson 4.4 Marine disasters 69

目录

前言

第 1 章　港口工程 ················· 73
1.1　港口规划与布置 ················· 73
1.2　码头功能分类 ··················· 74
1.3　码头结构分类 ··················· 75
1.4　荷载分类 ······················· 77
1.5　货物装卸设备 ··················· 78

第 2 章　航道工程 ················· 81
2.1　河道类型 ······················· 81
2.2　河床演变 ······················· 82
2.3　疏浚工程 ······················· 83
2.4　整治建筑物 ····················· 85
2.5　船闸 ··························· 86

第 3 章　海岸工程 ················· 88
3.1　海浪 ··························· 88
3.2　防波堤、海堤和丁坝 ············· 89
3.3　海岸地貌 ······················· 91
3.4　海岸输沙和海滩养护 ············· 92

第 4 章　海洋工程 ················· 94
4.1　海上平台 ······················· 94
4.2　波浪能 ························· 95
4.3　海底管线 ······················· 97
4.4　海洋灾害 ······················· 98

参考文献 ··························· 100

Chapter 1　Port Engineering

Lesson 1.1　Port planning and layout

Overall plan

Ports can be categorized into artificial ports and ports built in natural harbor. Artificial ports are built along the coastline by filling or digging. As shown in Figure 1(a), the land part of the port is built by filling soil. As shown in Figure 1(b), the port basin is artificially constructed by excavating the land near the coastline. The shape of the excavation basin depends on the size of the port and the method of excavation. The excavated port is connected to the sea through an approaching channel. At the entrance of the harbor basin, breakwaters are usually built to prevent the adverse effects of waves and currents.

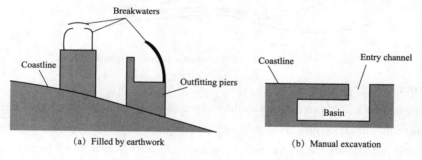

Figure 1　The layout of artificial ports

When selecting the port construction site based on the natural shoreline, important factors such as land availability, filling materials, soil quality, water depth, environmental conditions, etc., should be considered.

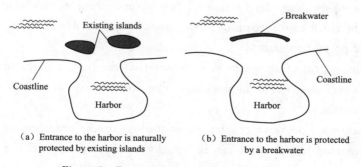

Figure 2　Ports constructed in natural harbors

Chapter 1 Port Engineering

In the overall planning of a harbor project, preliminary design standards should be proposed, covering designs such as containers, transshipment flows, import and export, ship type design, and operating equipment. The modern cargo port is more like a cargo handling hub in an intermodal transportation system than a maritime transport terminal. Therefore, the inland connection of the port is an essential element for the smooth operation and development of the port. The transfer of goods can be achieved through the connection of road or railway, artificial or natural inland waterway, airlines or petroleum product pipelines.

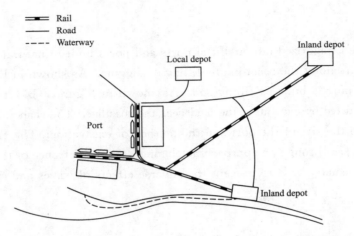

Figure 3 The connections between a port and inland depots

General layout

Port engineering arrangements should ensure easy berthing, safe cargo handling, and safe passenger disembarkation. It is necessary to design suitable navigation channels, port entrances, and port basin areas, as well as to avoid harmful erosion or deposition in and around the port area, to ensure that ships can easily enter the port.

When a ship enters the harbor basin, it needs to reduce its speed in order to make an anchoring action. Figure 4 is the layout of an artificial port with a maneuvering circle. The maneuvering area is outside the port, between the port entrance and the main port, or in the main harbor basin near the entrance. The maneuvering circle, also called turning waters, should be calculated according to the size of the port design ship.

General layout of protection works

The function of protection works is to ensure that the harbor basin and the surrounding area of the wharf are as calm as possible. The port protection project mainly includes the following buildings:

(1) Breakwaters – breakwaters can be connected to the coastline or detached. The breakwaters connected to the coast can be categorized into windward (or main breakwaters) and leeward (or secondary breakwaters). The windward protects the port from

Lesson 1.1 Port planning and layout

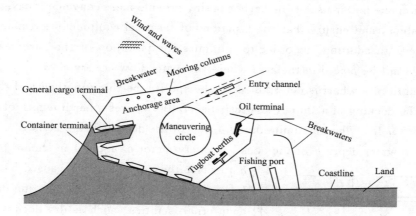

Figure 4 Layout of large multi-purpose artificial ports

waves in the main direction, and the leeward protects the port from waves in the secondary direction. Generally speaking, the leeward breakwater can be partially protected by the windward breakwater.

(2) Jetties – jetties are usually designed in pairs. They are arranged on the inward of coastline or in the river to regulate the entrance of the port. The paired jetties can also increase the flow velocity to prevent deposition.

Port engineering is usually located in the broken zone, where the sediment transportation is active. Generally, in port protection projects, sediment deposits on the upstream of the windward breakwater and erosion is found on the downstream of the leeward breakwater.

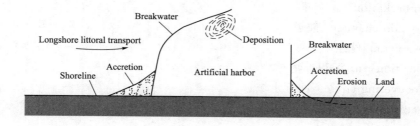

Figure 5 The impact of harbor engineering on coastline

General layout of inner harbor project

The rock and soil mechanical properties of the seabed in the project area have a great impact on the overall layout of the project. If there is a rocky seabed, it is usually recommended to place the dock front close to its design depth to avoid expensive cost of rock excavation. If the seabed is a soft soil foundation, detailed technical and economic analysis of reclamation and dredging must be carried out in the design of the wharf location.

Generally speaking, it must be assured in the layout planning that the shape of the terminal can make better use of the harbor basin, provide more convenient navigation conditions for ships, and ensure that the functions of terminal equipment and machinery are not hampered. In addition, in order to minimize the pollution of the harbor basin, the wharf should not be placed in the area where the water flow is very slow.

The length of a wharf is determined by the specific berthing method and the number of berth. The docking of a ship of length L generally requires a berth length of $b=L+30$ to 40m or $b=1.2L$. The maximum draught d_{max} of the designed ship is used to determine the minimum water depth h of the dock. A safety factor of about 1 m should be added to h to cover the heavy motions caused by waves, so $h \approx d_{max}+1$m. Except for the mooring docks, other internal facilities (such as dry docks, slipways, and maintenance docks) should be independent to the quay or wharf, and should be placed in protected port areas as much as possible.

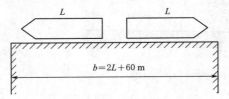

Figure 6 The length of a wharf

New words and expressions

1. port   n. 港口，口岸
2. overall plan   总体规划
3. filling   n. 填方
4. excavation   n. 挖掘，挖方
5. harbor basin   港池
6. hub   n. 中心
7. general layout   总平面布置
8. disembarkation   n. 登陆，上岸
9. navigation channel   航道
10. erosion   n. 侵蚀
11. deposition   n. 沉积（物）
12. maneuvering circle   回旋水域
13. breakwater   n. 防波堤
14. coastline   n. 海岸线
15. detach   v. 分离
16. windward   adj. 迎风的
17. leeward   adj. 背风的
18. jetty   n. 丁坝，突堤
19. sediment   n. 沉积物，泥沙
20. seabed   n. 海床
21. dock   n. 码头，船坞
22. reclamation   n. 填海造陆，围垦
23. dredging   n. 疏浚

24. draught   n. 吃水
25. dry dock   干船坞
26. slipway   n. 下水滑道
27. quay   n. 实体式顺岸码头
28. wharf   n. 透空式顺岸码头

Lesson 1.2 Classification of terminals by function

The functions of terminals are different depending on the modes involved and the cargo transferred. They can be classified into container, bulk and general cargo terminals.

Container terminal

Container terminals have low requirements for labor but have multiple intermodal functions. However, they need a lot of storage space, where containers can be handled by intermodal equipment (cranes, straddle carriers, racks, etc.), and because of the intermodal function of container terminals, special cranes, such as gantry cranes and container cranes, are also needed.

A container is a reusable box with different sizes, for example, the 20-foot standard container, the 40-foot standard container and the 40-foot high-cube container. The most commonly used type is the 20-foot long container, which is usually called the standard container. It also be used as a unit of measurement in the container transportation, that is, the Twenty-Foot Equivalent Unit (TEU). Normally, for a container ship, its capacity could be more than 1000 TEU. Containers can be used not only for container-fitted cargo, but also for liquid and refrigerated cargo.

Container terminal is a place for transferring containers between different means of transportation for further transport. Transshipment can be carried out between container ships and land vehicles, for example, trains or trucks. In this case, terminal is called a sea container terminal. Transshipment can also be carried out between land vehicles, generally, between trains and trucks. In this case, it is called an inland container terminal.

Container terminals are often a part of large harbor. They provide storage facilities for loading and unloading containers. The full container can be stored for a short period, waiting for continued transportation, while the empty container can be stored for a long period, waiting for the next load. Containers are usually stacked for storage in the area called container stack.

Bulk terminal

Bulk refers to a large quantity of unpackaged goods with the same size. Liquid bulk cargo includes crude oil and refined products, which can be transported by pumps, hoses and pipelines. Relatively limited handling equipment but a large number of storage facilities are required for liquid bulk cargo. Dry bulk cargo can be many kinds of products,

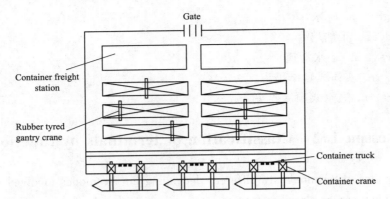

Figure 1　Container terminal

including ore, coal and grain. More handling equipment for dry bulk cargo is required, like specialized grab bucket, crane and belt conveyor system. For some specific bulk cargo, it may be necessary to make some changes to its characteristics to ensure the continuity of the transportation process, for example, the loading unit or physical state (from solid to liquid or gas, or any combination).

Bulk terminal is an industrial facility for the transfer of cargos, and storage of large quantities of cargos before they are transferred for processing or delivered to end users. Three main aspects are required to be considered in the design of the terminal: the products to be stored, the mode of transportation in and out of the facilities, and the throughput.

Bulk terminals usually consist of large pipelines, handling equipment, storage facilities and processing facilities. Let's look at several examples that bulk terminals provide critical infrastructure for various industries.

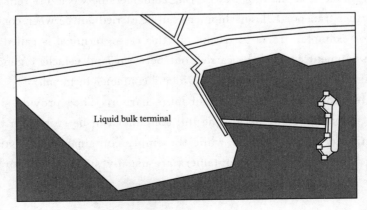

Figure 2　Liquid bulk terminal

Crude oil

Oil terminal is very important for the development of energy companies. Large oil tankers transport crude oil to the terminal. The oil is stored in tanks above or below the

Lesson 1.2 Classification of terminals by function

ground. Then, it is transported to the refinery. Terminals and refineries are usually built very closely to transport oil quickly for processing.

Grain

Bulk terminal is also important for the development of grain industry. Grain is usually transported to these terminals by ship, railway or truck, and then shipped out. These terminals have large storage capacity, and can transport grain more effectively through a combination of trains and inland river barges.

Cement

Bulk terminal is also important for the distribution of cement. The bulk terminal makes it possible to transport cement by barge, train or truck. These different ways of transportation provide flexibility to meet the needs of different customers. For example, barges are a cost-efficient way to transport large quantities of cement quickly. Trains could transport cement over a very long distance without worrying about traffic problems. Truck transportation provides fast and flexible choices for the growing markets.

General cargo terminal

General cargos refer to goods with various shapes, sizes and weights. Because the cargos are uneven and irregular, to carry them mechanically should be hard. General cargo handling usually requires manpower. Because of the variety of cargo forms, both storage yard and warehouse are required for a general cargo terminal.

With the development of the world's marine transportation, more and more cargos are transported by containers, which greatly squeezes the market of general cargo transportation. However, due to the special characteristics of general cargo themselves, container transportation cannot completely replace the general cargo transportation in the foreseeable future. Therefore, in order to build new type of general cargo terminals with low cost, high efficiency and high yielding, we should be creative in the preliminary planning and design of the terminals, and work out the most effective design scheme.

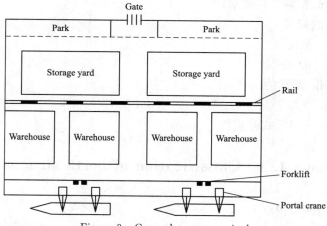

Figure 3 General cargo terminal

New words and expressions
1. terminal   n. 码头，中转站
2. general cargo   件杂货
3. container   n. 集装箱
4. intermodal   adj. 联运的
5. straddle carrier   跨运车
6. rack   n. 挂架
7. gantry crane   龙门式起重机
8. container crane   集装箱装卸桥
9. transshipment   n. 转运
10. harbor   n. 港，海港
11. stack   n. 堆垛
12. container truck   集装箱卡车
13. rubber tyred gantry crane   轮胎式龙门起重机
14. container freight station   拆装箱库
15. crude oil   原油
16. hose   n. 软管
17. pipeline   n. 管线
18. （dry/liquid) bulk cargo   （干/液体）散货
19. ore   n. 矿石
20. grab bucket   抓斗
21. crane   n. 起重机
22. belt conveyor   皮带运输机
23. throughput   n. 吞吐量
24. oil terminal   油码头
25. oil tanker   油轮
26. refinery   n. 精炼厂，炼油厂
27. barge   n. 驳船
28. cement   n. 水泥
29. warehouse   n. 仓库
30. marine   adj. 海洋的
31. forklift   n. 叉车
32. portal crane   门座式起重机

Lesson 1.3 Classification of berth structures

Wharf, quay, pier and jetty
  The main function of the berth structure is to provide a vertical front for safe bert-

Lesson 1.3 Classification of berth structures

hing. According to the layout of berth structures, they could be classified as wharf, quay, pier and jetty. Among them, wharf and quay are parallel to the shore while pier and jetty are extending out from the shore. Meanwhile, wharf and pier are generally built on piles while quay and jetty are solid constructions.

According to their respective construction style, the berth structures could be classified into several types. Three widely used types, i.e., gravity quay, sheet pile wharf and pile wharf, are introduced here.

Gravity quay

Gravity quay consists of the foundation, the wall body (gravity wall), the breast wall, the fender, the filter, and the back fill. Gravity quays are usually made of gravel or concrete in dry conditions, provided that the site can be drained and has a strong foundation. When the

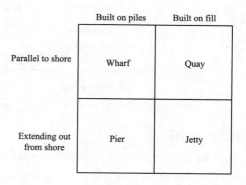

Figure 1 Types of berth structures according to their layout

foundation is relatively weak, gravity quays are built on piles in dry place. According to the different structural design, gravity quays can be subdivided into gravity block quay, caisson quay and so on.

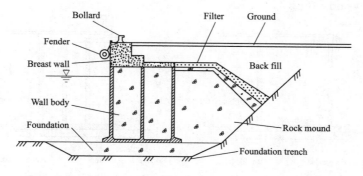

Figure 2 Gravity quay

Gravity block quay

The gravity block quay is one of the oldest types of gravity quay. It is made up of large blocks one after another, in a brick wall pattern. Such structures are made of high-quality natural stone or concrete blocks and are built on solid ground. They are durable structures that require only moderate maintenance. Because the natural stones are difficult to exploit, only concrete stones can be considered economically at present. As the erosion of pitching in front of the wall structure is a serious problem, many older block structures have to be protected with anti-erosion techniques.

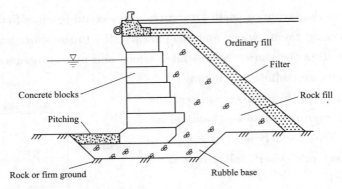

Figure 3 Gravity block quay

Caisson quay

For a caisson quay, the front part of the structure is built by arranging prefabricated concrete caissons in a row, which corresponds to the planned route of the new berth. The shape and design of caissons may be different, depending on the construction site and technique. Rectangular caisson is the most common caisson type. The caisson quay can relieve the stress on the outer edge of the caisson bottom more effectively than the block wall quay. Stress can be reduced by increasing the width of the caisson, or separating the caisson with two or three chambers while filling only the rear chambers. Caissons must also be designed to withstand the loads and stresses during their production, launching, towing, placement and filling.

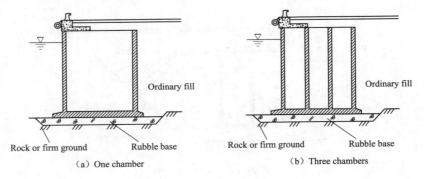

Figure 4 Caisson quay

Sheet pile wharf

Sheet pile wharf is a widely used type of berth structure in harbor engineering. It has been proved this structure is a feasible and economical type for waterfront buildings whose wall heights are no more than 18 to 20m, and whose site soil is suitable for piling. These walls are generally flexible. They are usually classified by the material of the sheet pile (for example, wood, steel, or concrete), the supports of the sheet pile, and the construction order of the wall. The main components of the sheet pile wharf include the sheet pile, the cap beam, the tie rod, the guide beam, the anchoring structure, and so on.

Lesson 1.3 Classification of berth structures

Sheet pile wharf is built by sinking sheet piles into the foundation. Its structure is simple thus reducing the use of raw materials. Also, the construction of sheet pile wharf is quick and convenient. However, the durability of the sheet pile structure is not as good as the gravity structure. During the construction of a sheet pile wharf, large wave force should be avoided.

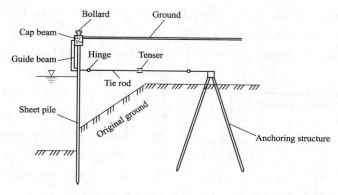

Figure 5 Sheet pile wharf

Pile wharf

In the small open-air berth structure with small water depth and limited load-bearing capacity of foundation, concrete or wooden piles are usually used for support. This structural style is called the pile wharf. If the platform needs to bear large loads, the piles must be placed closely together.

If the thickness of the foundation material is more than 3 to 4m, and if the ground treatment requires a large amount of divers and underwater works, the piles with much larger transections can be used. The piles could be concrete piles, steel pipe piles filled with concrete, or reinforced concrete piles. Compared with concrete piles, steel piles can be used for harder layers. For example, a steel pipe pile with a diameter of 70cm can penetrate a 20-meter-thick gravel fill made of stones up to 50cm in diameter. The pile with a diameter of 50 to 80cm is the most commonly used pile type.

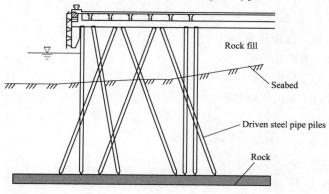

Figure 6 Pile wharf

Beam-slab type is a commonly used type for pile wharfs. In the first reinforced concrete wharf built in Norway, tall and narrow rectangular beams were used to support the slab. It was found that, after 10 to 15 years of use, these structures show a corrosion form of the steel bar at the bottom of the beam, as well as the subsequent cracking and spalling of the concrete covering the steel bar. In order to overcome these shortcomings, beamless platforms were invented, which had been proven to be very durable. However, such structures are more expensive to build, especially due to the need for template support systems. Thus, the beam-slab type structures are now used again. Modern beam-slab structures are not the same as the old ones. They have low beams and wide trapezoidal sections. Therefore, most of the shortcomings of the old beams are avoided. Nowadays, the trapezoid is a commonly used shape of the cross section of the beam in the pile wharf structure.

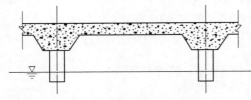

Figure 7   Cross section of a modern beam-slab structure

New words and expressions

1. berth   n. 泊位
2. pier   n. 栈桥式突码头
3. jetty   n. 实体式突码头
4. pile   n. 桩，v. 打桩
5. gravity quay   重力式码头
6. sheet pile wharf   板桩码头
7. pile wharf   高桩码头
8. breast wall   胸墙
9. fender   n. 护舷
10. filter   n. 滤层
11. back fill   回填
12. gravel   n. 砾石，碎石
13. concrete   n. 混凝土
14. caisson quay   沉箱码头
15. bollard   n. 系船柱
16. pitching   n. 铺石
17. prefabricated   adj. 预制的
18. stress   n. 应力
19. block wall quay   方块式码头
20. chamber   n. 室
21. launch   v. 下水
22. tow   v. 拖运

23. waterfront   adj. 滨水的
24. beam   n. 梁
25. cap beam   帽梁
26. tie rod   拉杆
27. guide beam   导梁
28. tenser   n. 紧张器
29. hinge   n. 铰链
30. concrete pile   混凝土桩
31. steel pipe pile   钢管桩
32. reinforced concrete   钢筋混凝土
33. slab   n. 板
34. spall   v. 裂成碎片
35. trapezoidal   adj. 梯形的
36. cross section   横截面

Lesson 1.4   Types of loads

One of the most important considerations when building a pier or wharf is the load that the structure must be able to withstand. Before construction, the load requirements of a given facility must be carefully analyzed to ensure that the expected traffic can be handled safely.

Dead load

The dead load should be considered primarily when designing a pier and wharf facility. The dead load is the weight of the entire structure, including all permanent accessories (such as light poles, public rooms, mooring hardware, utility pipelines, safes, sheds, and platforms). In order to plan the whole life of the facility, current and future accessories should be analyzed. When making these assessments, actual and available building material weights should be used.

The live loads and lateral loads are usually the control variable for the design of fixed piers and wharves. Therefore, overestimating the dead load usually does not have adverse impact on the total cost of the structure. Overestimating the dead load of floating piers and wharves, however, will result in an increase in costs.

The earth pressure may also affect the design of pier and wharf facility. Earth pressure may act on the fixed structure, causing the structure to move laterally.

Vertical live load

The vertical live (or moving/movable) load is a key factor for the construction of pier and wharf facility. The specific types of loads that occur on the structure will play an important role in design considerations.

The loading of railway cranes is common in pier and wharf facilities. Many cranes, like container cranes, have various capacities, configurations and gauges. Therefore, without specific information from the crane manufacturer, the final design should not be carried out. The increase in gauge will result in higher crane weight and higher wheel load. The maximum wheel load act on the deck plates, crane girders and pile caps should be increased by 20%.

On open and floating piers and wharves, truck cranes are commonly utilized. This tool has great limitations in operation. Therefore, the actual load should be specified according to the type of truck crane.

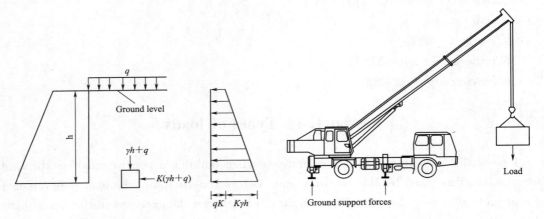

Figure 1   Earth pressure   Figure 2   Loads of a truck crane

Loads can be defined in several ways. For concentrated loads, like wheel loads and cantilever floating loads from tire vehicles (trucks, truck cranes, forklifts, etc.), the load should be applied at the position and in the direction that will cause the maximum force considered in the design. Generally, uniform and concentrated live loads should be applied in a logical manner to avoid simultaneous application of live loads from tires and railways in the same area.

It is important to consider concentrated load when designing short span structures such as decks and trench covers. This includes the loads from trucks, forklifts, mobile cranes and straddle trucks. Uniform load is a more important consideration in the design of beams, pile caps and supporting piles.

Horizontal live load

As important as the vertical live load, the designer must consider the horizontal live load of the piers and wharves before constructing the facility. The horizontal live loads play an important role in determining the type of elements used to disperse the forces acting on the structure and to protect the structure from the destruction over time.

For piers and wharves, the main function is the berthing and mooring of ships. The berthing load can be significant, especially when a ship is driven in with the help of two or more tugs. The fender system between a ship and the facility can reduce the berthing energy and the force transfered to the structure. The magnitude and location of the transmitted force depends on many factors, including the structure type, the vessel type, the approach speed and angle, and the fender system used.

Once the ship is moored, forces may be applied to the pier and wharf facility. These forces may be caused by winds, currents and waves. Many elements can determine the mooring load: the direction and strength of winds, currents and waves; the spacing of mooring points; the exposure of berth and the direction of ship; the layout of mooring line; the elasticity of mooring line; and the loading mode of ship. If the facility is in sheltered water, wave force is generally not considered. For all moored ships, however, currents and winds acting on them create forces. Computer simulation programs are often helpful in simulating potential mooring loads. When designing a pier or wharf structure, the mooring force should be considered during the construction process. In a multi-berth terminal, the structure itself should be able to hold the ship in the berth under maximum wind speed.

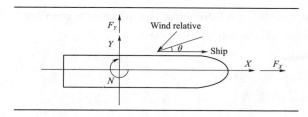

Figure 3 Wind load

Due to tidal fluctuation or groundwater accumulation, the water level difference may also generate pressure on facility elements. For example, on a sheet pile wharf, the water pressure will act on both sides of the sheet pile. The actual force on the sheet pile is the combination of them.

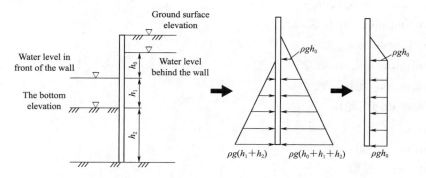

Figure 4 Water pressure on a sheet pile wharf

Other factors may also be related to the loads of pier and wharf structures, such as floating ice. Generally, compared with seawater ice of the same size and thickness, freshwater ice produces less pressure.

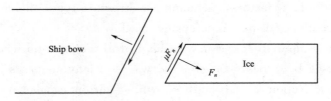

Figure 5　Ice load

Load combination

After considering each factor separately, the combination of various load types should be analyzed to construct wharves and dock facilities capable of withstanding the traffic on deck, the mooring of the vessel, and the forces on them. This requires a consideration of the combination of loads that may act on each component of the structure and foundation.

Determining the load requirements of a pier and wharf facility is a complex process that requires the use of computer models and predictions of the expected use of the facility. By using the best estimate and planning the maximum possible load, the designer can ensure a long service life of the wharf structure.

New words and expressions

1. dead load　固定荷载，恒载
2. permanent　adj. 永久的
3. pole　n. 杆
4. mooring　n. 系泊
5. safe　n. 保险库
6. shed　n. 棚
7. current　n. 流
8. live load　活荷载
9. fixed pier　固定式码头
10. floating pier　浮体式码头
11. earth pressure　土压力
12. configuration　n. 配置
13. gauge　n. 规格，尺寸
14. wheel load　轮压
15. deck　n. 平台，面板
16. girder　n. 梁
17. pile cap　桩帽
18. cantilever　n. 悬臂

19. mobile crane   移动式起重机
20. tug   n. 拖轮
21. elasticity   n. 弹力
22. shelter   v. 使遮蔽，n. 掩护物
23. bow   n. 船艏
24. load combination   荷载组合

Lesson 1.5 Cargo handling equipment

The nature of goods and the type of packing generally determine the selection of cargo handling equipment.

In a terminal or port, goods enter and exit ships, transit sheds, warehouses, barges, railways or road vehicles using various cargo handling equipment. The equipment includes two-wheeled trolleys and four-wheeled trucks. Some of them are manually propelled and some of them are driven by mechanical power. There are also mechanical or electrical tractors towing the four-wheeled trailers. There are also mechanically or electrically driven belt conveyors, extending from the quayside to the transit sheds, warehouses, railways or road vehicles.

There are all kinds of quayside cranes, horizontal luffing cranes, mobile cranes etc. on the docks to move and lift goods. Lifting equipment (lift-on /lift-off equipment) is also used to move the goods vertically.

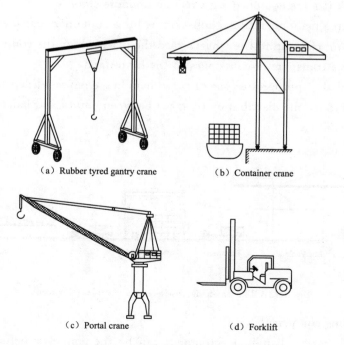

(a) Rubber tyred gantry crane   (b) Container crane

(c) Portal crane   (d) Forklift

Figure 1 Cargo handling equipments

Container handling equipment

One of the most common modes of transportation for non-bulk goods is container, as it can be moved seamlessly between ships, trucks and trains, simplifying the entire logistics chain. In fact, over the past few decades, the proportion of containerized cargo has been growing, reaching 90% of non-bulk cargo. Containerized cargo requires specialized handling equipment and carefully planned terminal configurations, to achieve the following objectives:

- Improve efficiency by ensuring that the right quantity is delivered timely at the right place in the most economical way.
- Minimize material damage during the transportation and storage.
- Optimize the storage configuration, maximize the use of space.
- Minimize cargo handling operation accidents.
- Reduce environmental emissions.

Container cranes, also called quayside container cranes, load and unload containers at the quayside. The tools used for moving and stacking containers at the terminal include transtainers, stacking cranes, container gantry cranes and container trucks. Containers are stacked up within the terminal up to five-tier high, which are, five containers per stack. The tools used for handling container cargos include forklifts and trucks, etc. Forklift trucks are usually driven mechanically or electrically. A platform in the shape of two prongs of a fork is installed in the front of the truck, with lifting capacity ranges from 1t to 45t. Some forklifts are equipped with reel and bundle clamps.

There are three main operational subsystems for a container terminal: the waterside-the water area from the ship to the wharf; yard-the place where the containers are stored; and intermodal-the transport of containers to the hinterland.

The choice of the appropriate type of cargo handling equipment depends on the size of the wharf, the layout, the distribution of space between empty and laden containers, and the operation flexibility.

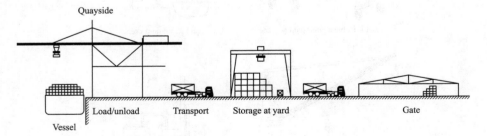

Figure 2   The handling process of a container terminal

Bulk cargo handling equipment

For dry bulk cargo, handling equipment can be the conveyor belts, usually with a

Lesson 1.5 Cargo handling equipment

hopper or a grab on the landward end, which may be magnetic to the ores, fixed on a large-capacity crane or gantry crane. These gantry frames move parallel to the dock, and also move a considerable distance landward, covering a large stacking area. This kind of equipment is used to process coal and ore. When handling bulk sugar, the sugar will be unloaded into a hopper, and delivered by gravity to railway trucks or road vehicles. Silos are usually related to grain. Pneumatic suction devices could suck grain out of a cabin.

The transportation of liquid bulk, like crude oil and refinery products, from the storage tanks, is carried out through pipelines connected to an onshore oil storage depot. The pumping equipment is installed on a tank storage device or a refinery ashore, but not on a quay. Due to the dangerous nature of such goods, the general practice is to build a special berth on the seaside near the main terminal system. The oil cargo is discharged from the oil tank of the ship, and it then goes through the pipeline system to the main manifold of the ship, which is usually located on either port or starboard side. And then through the loading arm on shore, the oil is transferred to the shore manifold, and finally distributed to the tanks on the oil terminal. The loading arm hose must be sealed to avoid oil leakage.

General cargo handling equipment

As for general cargos (merchandise, commodities, etc.), they are loaded and unloaded by the portal crane on the wharf, floating crane or by the ship's own handling equipment (deck cranes, derricks, etc.).

There are many types of tools or loose gears that can be attached to ship or shore lifting equipment. They include slings and rope straps, which are the common types of loose gears. This kind of equipment is generally made of ropes and is suitable for lifting strong packages, such as wooden cases or bagged goods, which are difficult to be sunken or damaged during lifting. Also, canvas is suitable for bagged goods. Sling is suitable for heavy, thin goods like wood or steel rails. Can hook or bucket hook is designed for lifting buckets or drums. Cargo net is for mail bags and similar goods, which are not easy to be crushed when lifting. Heavy lifting beam is designed for locomotives, railway passenger coaches and other heavy and long goods. Cargo tray and pallet, the latter of which is made of wood or steel, are ideal for medium-sized goods that can be conveniently stacked, such as cartons, bags, or small wooden crates.

New words and expressions

1. handling equipment 装卸设备
2. transit shed 中转仓库
3. trolley n. 手推车
4. trailer n. 拖车
5. quayside adj. 岸边的，码头前沿
6. horizontal luffing crane 水平变幅起重机
7. seamless adj. 无缝的

Chapter 1 Port Engineering

8. logistics chain 物流链
9. transtainer n. 移动式集装箱吊运车
10. stacking crane 堆垛起重机
11. prong n. 尖齿
12. reel n. 卷轴
13. clamp n. 夹子
14. yard n. 堆场
15. hinterland n. 内陆，内地
16. hopper n. 料斗
17. magnetic adj. 有磁力的
18. silo n. 筒仓
19. pneumatic adj. 气动的
20. onshore adj. 岸上的，陆上的
21. depot n. 仓库
22. ashore adj. 岸上的
23. manifold n. 总管，歧管
24. port side 左舷
25. starboard side 右舷
26. seal adj. 密封的
27. deck crane 甲板起重机
28. derrick n. 吊杆式起重机
29. loose gear 可拆卸零部件
30. sling n. 吊索
31. rope strap 绳带
32. sunken adj. 沉没的
33. canvas n. 帆布
34. hook n. 挂钩，吊钩
35. locomotive n. 火车头
36. cargo tray 吊货盘
37. cargo pallet 托货板
38. carton n. 纸板箱
39. crate n. 板条箱，篓

Chapter 2　Waterway Engineering

Lesson 2.1　Stream patterns

There are three basic types of stream patterns, straight, meandering and braided. Describing a river channel by the above terms does not mean that the entire river channel is straight or others. It just means certain reaches of the river channel can be described in this way. In fact, some parts of the river may be straight, some may be meandering and/or braided.

Straight stream

A straight stream is generally considered as one of the typical plain channel landforms in conventional classification. The straight stream is mainly unstable and develops along the faults and joints. For self-adjusted fluvial rivers, very few straight rivers are distributed in a larger spatial and temporal span. Therefore, the questions arise: could a stream be straight? What are the main factors that control the formation of streams and the transformation from the straight stream to another pattern? Various hypotheses and theories, such as hypothesis of geomorphic threshold, hypothesis of energy dissipation extremum, and theory of stability can answer the questions aforementioned, but they cannot explain the formation of straight stream. In view of the modern fluvial plain morphology, the straight stream is not as stable as other typical stream patterns in nature. From the historic records of the river sedimentation, no strong evidence could be found to support the stable evolution of a straight stream.

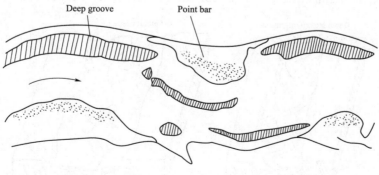

Figure 1　Straight stream

Although few river channels are completely straight, it is easy to describe a river channel as a straight one. A meandering channel is one that takes twists and turns over its length. Geoscientists use the sinuosity to determine whether a channel is straight or meandering. The sinuosity refers to the ratio of the actual channel length to the straight line distance between two points. A river is said to be meandering when its sinuosity ratio exceeds 1.5.

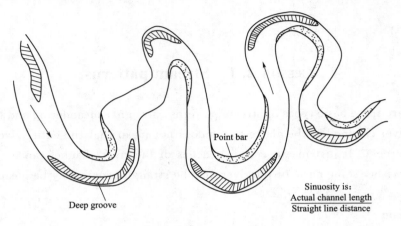

Figure 2 Meandering stream

Meandering stream

The river meanders refer to a series of curves, bends, turns or loops in river channels, streams or other watercourses. It occurs in the process of the river or channel swinging or shifting from one side to the other as it flows across the floodplain or river valley. River meanders are caused by the erosion of the concave banks and silting up on the convex banks downstream. As a result, the river channels migrate along the central axis to the lower reaches of the floodplain forming a curved channel. A meander belt refers to the area where the meandering stream often diverts and swings at the floodplain or the bottom of the valley, usually 15 to 18 times the width of the river channel. Meandering streams move downstream over time, and sometimes change significantly in a short period of time, causing municipal engineering problems for the maintenance of roads and bridges.

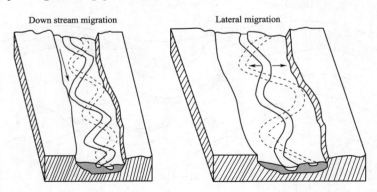

Figure 3 Fluvial processes of a meandering stream

Lesson 2.1 Stream patterns

Meandering stream is formed under the influence of natural conditions and processes. The wave structure of a river is continuously changing. Current flows in the form of vortex in the bend. Once the river begins to flow in sinusoidal path, the helicoidal flow will transport a large amount of sediment to the convex bank of the bend, while the concave bank of the bend is not protected and therefore vulnerable to serious erosion. Thus the amplitude and concavity of the meanders increase dramatically and finally form a positive feedback cycle.

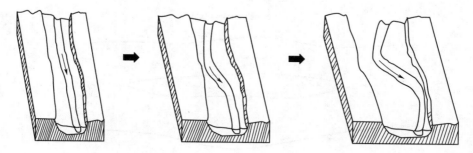

Figure 4　Formation of a meandering stream

The transverse current at the bottom of a stream carries a large amount of sediment to the interior of bend. After that, the transverse current flows upwards to the surface near the convex bank of the river, and then flows to the concave bank to form a helicoidal flow. The greater velocity of flow and curvature of river channel, the more intense the transverse current and erosion will be.

On the premise of conservation of angular momentum, the velocity of convex bank side is faster than that of concave bank side. Since the velocity of the flow decreases, so does the centrifugal force. In this case, the pressure of the higher water column is dominant, creating an unbalanced gradient that moves water back across the bottom from the concave bank side to the convex bank side. This secondary flow carries sediment from the concave to the convex bank side of the bend, gradually making the river meander.

Braided stream

A braided stream is composed of intertwined channels separated by one or several middle stream islands. Braided streams are often found in rivers with high sediment concentration and coarse bed material, as well as in situations where the slope of river is steeper than typical straight and meandering streams, which is also related to the rapid and frequent variation in the amount of water and the vulnerability of river banks to erosion. Braided streams are distributed in various environments around the world, such as mountain rivers dominated by gravel, and sandy bed rivers flowing across alluvial fans, sedimentary plains or estuarine deltas.

Braided channels are usually found in reaches where the banks consist of easily erodible sandy material with little vegetal protection. Bed material is relatively coarse and heterogeneous. The slope of the braided reach is greater than that of adjacent unbraided reaches. Hydraulically, the braided reach is less efficient than the unbraided reach. The total width of branches in a braided reach may be 1.5 to 2 times that of an undivided channel, and the depth of flow is correspondingly less. Braiding is thus a way of dissipating energy when stream slope steepens. Velocity increases that would otherwise lead to erosion are thus avoided.

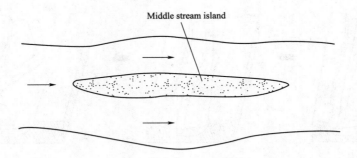

Figure 5   Braided stream

New words and expressions

1. meandering   adj. 弯曲的，曲折的

2. braided   adj. 辫状的，分汊的

3. reach   n. 河段

4. fault   n. 断层

5. joint   n. 节理

6. fluvial   adj. 河流的，河流冲刷形成的

7. hypothesis of geomorphic threshold   地貌阈值假说

8. hypothesis of energy dissipation extremum   能量耗散极值假说

9. fluvial plain   冲积平原

10. sinuosity   n. 蜿蜒度

11. point bar   边滩

12. groove   n. 深槽

13. floodplain   n. 河漫滩，滩地

14. concave bank   凹岸

15. silt   v. 淤积

16. convex bank   凸岸

17. vortex   n. 涡流，旋涡

18. helicoidal flow   螺旋流

19. amplitude   n. 振幅，幅度

20. transverse   adj. 横向的

21. conservation of angular momentum 角动量守恒
22. centrifugal force 离心力
23. intertwined channel 交织水道
24. middle stream island 江心洲
25. sediment concentration 含沙量
26. bed material 河床质
27. alluvial fan 冲积扇
28. estuarine adj. 河口的
29. delta n. 三角洲
30. heterogeneous adj. 由很多种类组成的，不均匀的

Lesson 2.2　Fluvial processes

Fluvial processes refer to the process of erosion and silting of the riverbed under natural conditions and after the construction of regulating structures, including sediment movement and erosion or deposition of the riverbed.

Fluvial processes of alluvial river in flatland are mainly manifested in the development and change of the sediment bodies in the channel. The change of sediment bodies is not only the increase of siltation and the decrease of erosion, but also the plain displacement causing the change of channel shape. The river bank will be eroded back in some places and silted in other places.

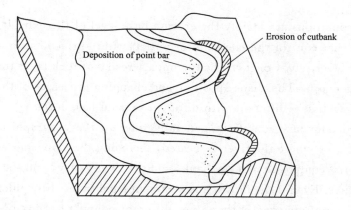

Figure 1　Erosion and deposition along a meandering stream

The fluvial process of the plain river is closely related to the stream patterns. Different stream patterns have different evolution rules and formation conditions. The sediment body of straight and meandering river reaches are mainly point bars. Among them the point bars of straight reaches are distributed indented on both sides of the channel, while the point bars of meandering river reaches are attached to convex banks. There are central islands or bars in braided reaches and wandering reaches, among which braided reaches have stable river central islands, while the wandering reaches have many unstable river central bars which tend to migrate frequently. The form of an alluvial stream depends largely on the relative erodibility of the bed and banks.

An initially straight channel, either in a laboratory flume or in the field, will usually develop meanders as water flows through it if the bank material is erodible. A meandering channel may be 1.5 to 2 times as long as a non-meandering channel. Its slope is correspondingly reduced, but head losses are increased due to both the longer channel length and the bend losses. Without these losses, velocities would be higher, with corresponding tendency to downcut the channel. Many meandering streams cannot downcut because they discharge into a water body with fixed elevation. If downcutting cannot occur, some other mechanisms are required to dissipate the available energy.

Thus both braiding and meandering can be explained as means of energy dissipation. Braiding will occur when bed material is coarse and heterogeneous and banks are easily erodible. Meandering is likely to occur on flatter slopes where the material is finer and the banks somewhat more cohesive.

The main factors affecting fluvial processes

The main factors affecting fluvial processes can be summarized into three aspects: inlet condition, outlet condition and riverbed boundary condition.

The inlet conditions include the rate of volume of inflow water and its change process, the oncoming load and its composition and change process, and the connection mode with the upper reach.

The outlet conditions are mainly the erosion base level of the exit. It can be all kinds of water surface that control the elevation of outlet water surface, such as river surface, lake surface, sea level, etc., or it can be the impact-resistant rock layer that limit the river to downcut in the depth. The shape, position and changing processes of the river's longitudinal profile are obviously different with different erosion base level.

The riverbed boundary conditions generally refer to the geographical and geological conditions of the area where the river is located, including the slope and width of the river valley, the material composition of the banks and the river valley, and the geometric shape of the river channel. Even if the inflow and sediment conditions of the inlet and the erosion base level of the outlet are exactly the same, different boundary conditions of the riverbed will still bring about different characteristics of fluvial processes.

The essential reason for fluvial processes

For a certain reach of any river, the riverbed will be deformed by erosion and silting when the amount of sediment flowing in and out of this reach is not equal. If the amount of sediment entering the area is greater than the current in the area can carry, the bed will be silted up. Conversely, if the amount of sediment entering the area is less than the current in the area can carry, the riverbed will be eroded.

Therefore, although the specific reasons for the fluvial process vary greatly, they can be fundamentally attributed to the result of the unbalance of sediment transport, that is, the fluvial process is the direct consequence of the unbalanced sediment transport.

When the external conditions, i. e., the water and sediment conditions of inlet, the erosion base level conditions of outlet, and the riverbed boundary conditions remain constant, the whole reach is in a state of equilibrium sediment transport, each part of the reach may still be in non-equilibrium. For example, the existences of sand waves and sediment bodies make the near-bottom flow accelerate and decelerate alternately. Erosion occurs in the flow accelerating zone and siltation occurs in the flow decelerating zone. As a result, the riverbed, which is still in a state of equilibrium sediment transport, is in an unbalanced state at particular spots.

Another important reason for the non-equilibrium state of sediment transport is the constant variation of inlet and outlet stream conditions. Due to the heterogeneous spatial and temporal distribution of precipitation in the basin, the conditions of flow and sediment of inlet almost always change. As for the outlet conditions, if the objective is the erosion base level mentioned before, the change is very slow; if the focus is on the change of flow conditions, such as the mutual support of main stream and tributaries, or the impact of tidal waves on floods, it may change a lot. The riverbed boundary conditions are usually relatively stable, but when the boundary is deformed, such as during and after a waterway regulation project, a new non-equilibrium of sediment transport may be triggered.

New words and expressions

1. fluvial process   河床演变
2. alluvial river   冲积河流
3. siltation   n. 淤积，聚集
4. cutbank   n. 陡岸
5. indent   v. 使成锯齿状，使犬牙交错
6. erodibility   n. 易蚀性
7. laboratory   n. 实验室
8. flume   n. 水槽
9. head loss   水头损失
10. downcut   v./n. 下切侵蚀
11. elevation   n. 高程
12. cohesive   adj. 有黏着力的
13. erosion base level   侵蚀基准面
14. longitudinal profile   纵剖面
15. geographical   adj. 地理学的
16. geological   adj. 地质的

17. equilibrium   n. 平衡，均衡
18. sand wave   沙波
19. accelerate   v. 加速
20. decelerate   v. 减速
21. precipitation   n. 降水
22. basin   n. 流域，盆地
23. tributary   n. 支流
24. tidal wave   潮波
25. waterway regulation   航道整治
26. trigger   v. 触发，引起

Lesson 2.3 Dredging engineering

Dredging is a form of engineering in which submerged material is excavated and moved from one place to another in water or out of water with equipment called dredgers. The purpose of dredging is to maintain waterways and harbors navigable, and to assist in coastal protection, land reclamation and coast redevelopment by grabbing and transporting bed materials to other locations. Dredging can also recycle materials with commercial value, including certain minerals or sediments that can be used as building materials, such as sand and gravel.

Dredging works are usually classified into three categories: capital, maintenance and temporary projects. Dredging in new sites or in materials that have never been dredged before is called capital dredging projects, such as reclamation construction of airports and artificial islands, development of new ports, deepening and widening waterways. Maintenance dredging refers to regular dredging works to maintain or improve existing waterways. Temporary dredging works are to solve the dredging tasks of small quantities. It is usually carried out on a reach where there is no permanent dredger, temporarily using the dredging equipment from adjacent area.

Dredging usually involves four steps: loosening the bed material, moving it to the surface, transporting and disposing. According to the working principle, dredgers can be divided into two types which are hydraulic dredgers and mechanical dredgers.

Hydraulic dredger

The key characteristic of a hydraulic dredger is that the working material is suspended, and raised through the pumping system and fed to outlet pipes. Gravel and other coarse sediments can also be removed by a more powerful hydraulic dredger. Some common hydraulic dredgers are briefly described below:

Plain suction dredger（SD）

When working, a plain suction dredger can be fixed in position by using one or more spuds. Plain suction dredgers use floating pipes to transport sediment to the shore where it needs to be filled. Long distance delivery requires an additional booster pump in the pipeline. Sediment can also be loaded directly into barges moored alongside.

Cutter suction dredger (CSD)

The inlet of the suction tube of a CSD is equipped with a cutter device which makes the bed material loose then transports it to the suction inlet. The dredge sediment is sucked out by wear-resistant centrifugal pumps and discharged through pipes or barges. A cutter suction dredger is often used in areas with tough bed material, such as gravel deposits or surface bedrock. With sufficient power, a cutter suction dredger can be used instead of underwater blasting.

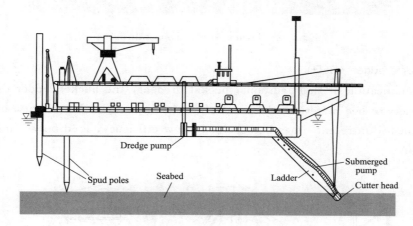

Figure 1　Cutter suction dredger

Trailing suction hopper dredger (TSHD)

TSHD is a self-propelled ship, which fills its cabin or hopper according to a preset track in the dredging process. The hopper can be dumped at the bottom or through a valve, or the sediment can be loaded onto the shore through a mud pump. This type of dredger is mainly used in open waters, such as rivers, canals, estuaries and open seas.

The appearance of a TSHD is the shape of a conventional ship, which has good seaworthiness and can be operated without any form of mooring or spud. The standard used to measure a TSHD is the capacity of hopper which can vary from hundreds of cubic meters to over 20000m^3. The construction of larger vessels in recent years has made the transport of dredged materials more economical, particularly for reclamation projects.

Mechanical dredger

Mechanical dredgers include a variety of forms, and each has the same "grab" working principle. These devices are equipped with a grab or bucket to pick up loose bed sediment, and fill the bucket with material, then lift the bucket to transport it to the dump site. Some common mechanical dredgers are briefed as follows:

Chapter 2 Waterway Engineering

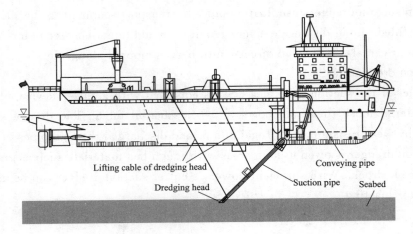

Figure 2 Trailing suction hopper dredger

Bucket ladder dredger (BLD)

As a modification of the traditional bucket dredger, the bucket ladder dredger can handle a variety of bed materials, such as soft rocks and corals. But the use of BLD has hugely diminished in recent times because of its low efficiency, loud noise, and the need for anchor line.

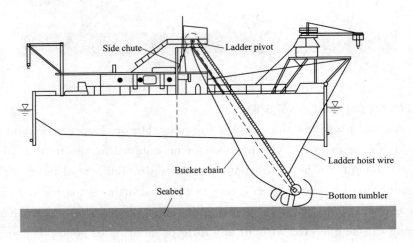

Figure 3 Bucket ladder dredger

Backhoe dredger (BHD)

A bucket dredger has a backhoe like an excavator. A backhoe excavator for ground use equipped on a floating dock or barge can be a simple but usable backhoe dredger. A pint-sized backhoe dredger can be installed on a track near the trench shore for operation. Dredged materials are usually carried on barges. Backhoe dredgers are mainly used in harbor areas and other shallow water areas.

Lesson 2.3   Dredging engineering

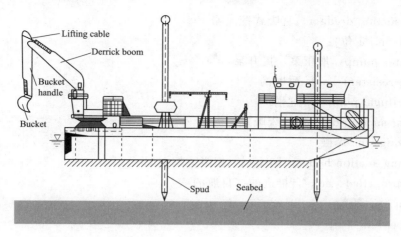

Figure 4   Backhoe dredger

Grab dredger (GD)

A grab dredger uses a clamshell bucket on a ship-borne crane or a floating crane to grab the bottom sediments. This device is commonly used in bay mud excavation. Most of these dredges are crane barges equipped with spuds and lifting steel piles.

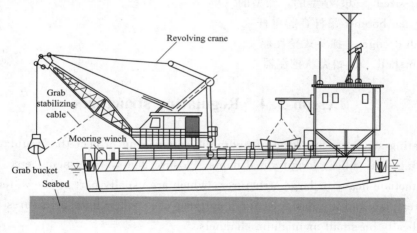

Figure 5   Grab dredger

New words and expressions

1. dredger   n. 挖泥船
2. navigable   adj. 可航行的，适于航行的
3. land reclamation   填海造陆
4. mineral   n./adj. 矿物质（的）
5. artificial island   人工岛
6. temporary   adj. 临时的，暂时的
7. hydraulic dredger   水力式挖泥船
8. mechanical dredger   机械式挖泥船

9. plain suction dredger   直吸式挖泥船
10. spud   n. 定位桩
11. booster pump   增压泵，提升泵
12. wear-resistant   adj. 耐磨的
13. centrifugal pump   离心泵
14. cutter suction dredger   绞吸式挖泥船
15. blasting   n. 爆破
16. trailing suction hopper dredger   耙吸式挖泥船
17. self-propelled   adj. 自推进的，自航的
18. cabin   n. 船舱
19. hopper   n. 料斗
20. canal   n. 运河
21. seaworthiness   n. 适航性
22. bucket ladder dredger   链斗式挖泥船
23. anchor line   锚线
24. backhoe dredger   铲斗式挖泥船
25. pint-sized   adj. 小型的，微型的
26. derrick boom   吊杆，起重杆
27. grab dredger   抓斗式挖泥船
28. clamshell   n. 蛤壳状挖泥器

Lesson 2.4 Regulating structures

Regulating structure is a kind of hydraulic engineering project that regulates a channel, including spur dikes, longitudinal dams, closure dams, revetment, etc. Regulating structures include heavy or large structures, which are usually part of the waterway regulation master plan and are designed for long-term use, while light structures are mainly used periodically on small or medium channels.

Spur dike

A spur dike is a structure protruding from the shore to protect the embankment from erosion. Such structures are widely used in waterway regulation engineering project with one or more of the following functions:

(1) Spur dikes guide the river along the desired course by diverting, deflecting and maintaining the flow in the channel.

(2) Spur dikes can form a slow flow zone to promote the siltation of the area near the spur dikes.

(3) Spur dikes prevent the bank from eroding by the main current and protect the embankment.

Lesson 2.4 Regulating structures

According to their purposes, spur dikes can be used individually or in combination. The spur dike may be perpendicular to the main flow, or may point upstream or downstream at a certain angle. They can also be used in conjunction with other regulation measures. If a river section to be protected is long, or if a single spur dike is insufficient to effectively alter the flow, and thus its impact on upstream and downstream sediments is not obvious enough, spur dike groups can be used. In a group of spur dikes, the upstream one is more vulnerable to the impact of current at both its river end and land end. Special treatment should therefore be given to ensure the stability of structures.

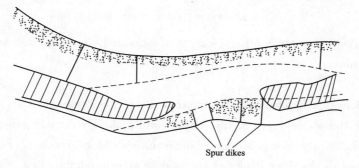

Figure 1　A group of spur dikes to fix the point bar

By acting on the surrounding water flow, a spur dike tends to increase the local velocity and turbulence intensity near it. The structure of the spur dike itself may be susceptible to erosion. The water flowing parallel to river banks is resisted by the spur dike, then accelerates along the upstream side of the spur to the spur dike head. The increase of flow velocity and curvature near the head of spur dike may cause severe erosion of the riverbed nearby. The regulating structure must have a sufficiently deep foundation or have sufficiently good protection, otherwise the head and root of spur dike may be damaged by local scour.

The angle between the spur dike and the main flow may affect the result of regulation. The spur dike perpendicular to the river is usually the shortest and therefore the most economical. An upward spur dike provides better protection of the river end of the structure itself from scouring. A downward spur dike is suitable for the protection of a concave bank, especially if the length and spacing of the spur dikes keeps the mainstream away from the concave bank, thus providing protection for the entire reach.

Longitudinal dam

A longitudinal dam is a regulating structure with longitudinal layout. The body of the dam is generally long, and approximately parallel to the direction of the flow or with a small intersection angle. It is usually arranged along the regulation line. The functions of a longitudinal dam include narrowing the channel, guiding the water flow and adjusting the shoreline, therefore, it is also called a diversion dam. The longitudinal dams are usu-

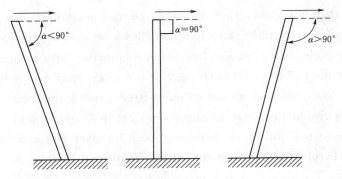

Figure 2　Angle between the spur dike and the flow

ally used at the diversion and confluence areas of a braided stream as well as the regulating reaches of an estuary.

Closure dam

A closure dam is a hydraulic structure that intercepts the anabranch of the river. In order to increase the water depth of the channel and make a river navigable, some narrow river channels that are unnavigable and dangerous should be blocked. Both ends of the dam are embedded in the river banks or the middle stream island. The central part of the dam crest is horizontal, and its two sides rise to the river banks.

Ecological revetment

To avoid the adverse effect of flood and erosion, concrete and rock materials have been used extensively in traditional embankment revetment projects. Firm revetments and stable structures are crucial for ensuring the safety of people and their properties. But solid revetments have negative effects on the ecosystem, including habitats of aquatic and amphibian, water quality of river as well as aesthetic values. The ecological effects of bank revetment projects must be considered in order to build safe bank revetments while reducing the negative impact on river ecosystem.

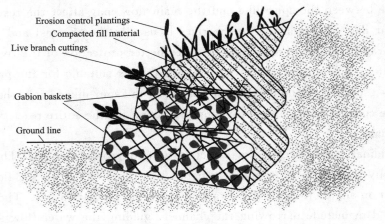

Figure 3　Ecological revetment structure

Lesson 2.4　Regulating structures

Ecological revetment work must combine civil engineering technology with topography, hydrology, ecology and other conditions of river to ensure the stability of banks and effectiveness of ecological restoration. Porous structures and plants, an indispensable part in the concept of "Sponge City", are often used in ecological revetments to promote the circulation of groundwater and river water, and facilitate the ecological restoration of river banks. Limestone materials can also be used for the construction of ecological revetments. Porous structures for the ecological revetment include rock slope protection, gabion protection, arc stone bed protection, etc., which provide habitats for the reproduction of microorganisms and plants, and have vital function for treating water pollution. The growth and diversity of plants and microorganisms have a positive effect on the quality of river water.

New words and expressions

1. regulating structure　整治建筑物
2. spur dike　丁坝
3. longitudinal dam　顺坝
4. closure dam　锁坝
5. revetment　n. 护岸
6. master plan　总体规划，蓝图
7. protrude　v. 伸出，凸出
8. embankment　n. 堤防
9. perpendicular　adj. 垂直的
10. turbulence　n. 湍流，紊流
11. intersection angle　交角
12. regulation line　整治线
13. diversion dam　导流坝
14. confluence area　汇流区
15. estuary　n. 河口
16. intercept　v./n. 拦截
17. anabranch　n. 支流，重汇支流
18. habitat　n. 栖息地
19. aquatic　n./adj. 水生植物或动物（的）
20. amphibian　n. 两栖类
21. aesthetic　adj. 审美的，美学的
22. topography　n. 地形，地貌
23. porous　adj. 多孔的，渗水的
24. Sponge City　海绵城市
25. limestone　n. 石灰岩
26. gabion　n. 填石铁笼

27. microorganism　n. 微生物
28. diversity　n. 多样性

Lesson 2.5　Locks

A ship lock is an infrastructure that facilitates channel navigation. Due to the influences of flow regulation and channelization works in natural rivers, as well as the restrictions of bed topography and water surface slope in the canals, it is frequently necessary to create a streamwise stepped water level profile. Therefore, a specific navigable structure must be used to help ships pass through such a water level drop smoothly. The most commonly used modern navigable structure is the ship lock. It is composed of upper and lower approach channels, upper and lower lock heads, and a lock chamber. The lock chamber that can anchor ships or fleets is a box-shaped chamber, where the water level is adjusted by filling and emptying water, thus the ships can rise or fall vertically between different elevations to pass through the abrupt water level drop in the waterway. When ships or fleets travel from downstream to upstream of the lock, the water level in chamber drops to the same level as the downstream side. After the gate of downstream lock head is opened, ships or fleets enter the chamber, then the gate is closed and the chamber is filled with water. When the water level rises to the elevation of the upstream side, the ships can sail upstream through the upper approach channel when the gate of upstream lock head is opened. When ships or fleets move downstream, the procedure of ship lock control is reversed.

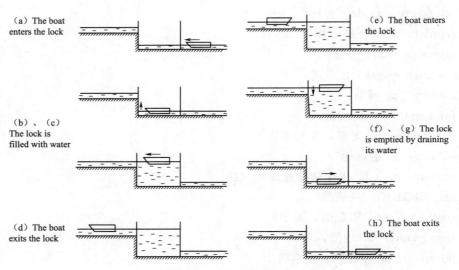

Figure 1　Operation of a lock

The most important characteristic of a ship lock is that there is a fixed lock chamber, in which the water level can rise and fall, whereas a caisson lock or ship lift can move up and

Lesson 2.5 Locks

down itself (usually called a caisson). A ship lock consists of a rectangular lock chamber with fixed sides, removable gates and facilities for water filling and emptying. Water filling and emptying in the chamber can be achieved by sluices operated manually or mechanically. The size of the chamber depends on the maximum possible size of the vessel using the waterway. In heavy traffic channels, two or more chambers may be required.

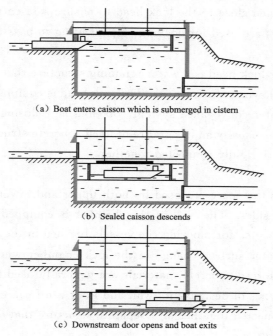

(a) Boat enters caisson which is submerged in cistern

(b) Sealed caisson descends

(c) Downstream door opens and boat exits

Figure 2　Operation of a caisson lock

Main components of a ship lock

A ship lock is composed of lock heads, lock chambers, water conveyance system, operating gates, valves, approach channels and corresponding equipments.

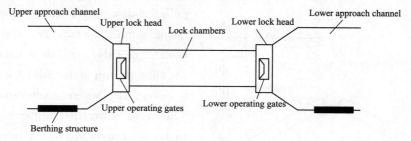

Figure 3　Main components of a ship lock

Approach channel

The approach channels, which are divided into the upstream approach channel and the downstream approach channel, are transitional sections connecting the lock and the main channel. The plane shape, width, and depth of the approach channel should enable ships

37

or fleets to enter and exit the lock chamber safely and rapidly. The direction and velocity of water flow at the inlet and outlet of the approach channel should be able to meet the requirements of safe entry and exit of vessels, as well as to prevent sediment from being deposited in the approach channel due to backflow. Generally, there are navigation and berthing structures in the approach channel. Navigation structures are impermeable navigation walls, which are arranged close to the lock head to ensure safe entry and exit of vessels. The berthing structures are used for docking ships waiting to pass the lock.

Lock head

An upper or lower lock head is a water retaining structure that separates the chamber from the upstream or downstream channel. The lock head is equipped with working gates, maintenance gates, water conveyance system, opening and closing devices for gates and valves, etc. The head is usually an integral reinforced concrete structure, whose abutment pier and bottom slab are rigidly connected together.

Lock chamber

A lock chamber is the space enclosed by both upper and lower lock heads as well as the lock walls on two sides. The wall of the chamber is equipped with mooring pillars, floating mooring rings, etc. for mooring the vessels berthed in the lock chamber. Vessels rise or fall with the water surface change in the lock chamber when passing through the lock. Masonry or reinforced concrete materials are generally used for the chamber structure. There are two types of designs: its wall and floor can form either an integral structure that is rigidly connected together or a separate structure that is not connected.

Operating gate and valve

An operating gate is a movable water retaining device installed on the upper or lower lock head. The gate is opened or closed when water levels on both sides of the lock head are the same. Because of its frequent operation, the gate is required to operate flexibly and quickly. The miter gate is the most widely used gate type while plate lift gate, horizontal pulled gate, sector gate (also known as triangle gate), etc., are also commonly used. The valve, which is used to control the flow during water filling or emptying, is set in the water conveyance corridor. The valve is opened under the action of hydraulic pressure, which requires simple structure, small hoisting force and convenient operation. Three types of valves, namely plate lifting valve, butterfly valve and reverse arc valve are the most commonly used ones.

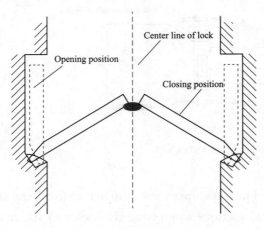

Figure 4 A plan view of a miter gate

Lesson 2.5 Locks

Water conveyance system

The water conveyance system is a facility to fill and empty water into/from the lock chamber. The filling or emptying time of the chamber should be as short as possible, and meet the requirement of stable vessel berthing, which is generally 10-15 minutes. There are two basic forms of water conveyance system:

1) Centralized water conveyance system (also known as the water conveyance system at the head). The filling and emptying of the lock chamber are carried out at the lock head through the water conveying corridor in the upper and lower lock heads, respectively.

2) Disperse water conveyance system. The filling and emptying water of the lock chamber is carried out through the water conveying corridor and the water outlet distributed in the bottom slabs or the chamber walls. If a water head (the maximum water level drop between the upstream and the downstream sides of the lock) is within 15m, the centralized water conveyance system is generally adopted. If the water head is larger, disperse water conveyance system should be used instead.

New words and expressions

1. channelization   n. 渠化
2. approach channel   引航道
3. lock head   闸首
4. lock chamber   闸室
5. anchoring   n. 锚定
6. caisson lock   箱形船闸
7. ship lift   升船机
8. sluice   n. 泄水闸
9. cistern   n. 水箱，蓄水池
10. backflow   n. 回流
11. berthing structure   靠船建筑物
12. abutment pier   边墩
13. bottom slab   底板
14. mooring pillar   系船柱
15. masonry   n. 砌石
16. miter gate   人字形闸门
17. plate lift gate   平板升降闸门
18. horizontal pulled gate   横拉闸门
19. sector gate   扇形闸门
20. water conveyance corridor   输水廊道
21. hoisting force   启闭力
22. centralized water conveyance system   集中输水系统
23. disperse water conveyance system   分散输水系统
24. water head   水头

Chapter 3 Coastal Engineering

Lesson 3.1 Ocean waves

Wave generation

Ocean waves are primarily produced by wind force exerted on water. Waves are initially generated by a complicated process of shear and resonance, during which waves with different wave heights, wave lengths and wave periods are formed and propagated in different directions. Once formed, waves can travel over a long distance with decreasing wave height, but the wave length and period can remain unchanged during propagation.

In the generation zone of storm, the energy of higher frequency waves (i.e., short period waves) is damped or transferred to that of lower frequency waves. Waves of different frequencies propagate with different speeds. Thus the sea state is changed by the separation of various frequency components outside the storm generation area. Low-frequency waves travel much faster than high-frequency ones, leading to swells instead of wind seas. This process is called dispersion effect. Therefore, wind waves are characterized as steep, irregular, and short-crested waves including a series of frequencies and directions. In contrast, swell waves are characterized as relatively mild, rather regular, and long-crested waves comprising a narrow range of frequencies and directions.

Wave transformation and attenuation

Wave shoaling

In fluid mechanics, wave shoaling refers to an effect of wave height change when waves enter shallower water. This is because the wave group velocity, that is, the transmission speed of wave energy, varies with the water depth. In a steady state, it is necessary to increase the energy density to compensate for the decrease of transmission speed in order to keep a constant energy flux. The wave length of a shoaling wave decreases but its frequency remains unaltered.

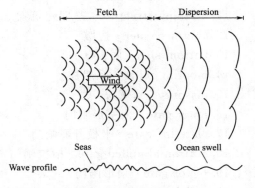

Figure 1 Wave generation and dispersion

Seabed friction

Waves in both transitional and shallow waters can be attenuated by the dissipation of

Lesson 3.1 Ocean waves

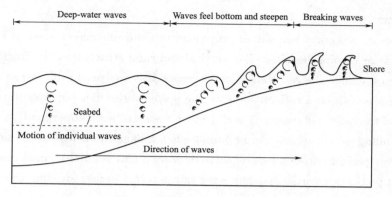

Figure 2 Wave shoaling process

wave energy from the bottom friction. This energy loss can be calculated approximately by the linear wave theory, which is identical to the friction relationships for pipeline and open channel flow. Different from the velocity distribution in a steady flow, the friction effect associated with waves can produce a very small oscillatory wave boundary layer. Therefore, the velocity gradient is significantly higher than that in an equivalent steady flow. Accordingly, it means that the coefficient of wave friction can be several times larger.

Wave refraction

It is a process in which the direction of a traveling wave changes due to its interaction with the seabed topography. Waves travel thousands of kilometers in the open sea before they reach the shoreline. Usually, they reach the shoreline perpendicular to the beach, generating waves parallel to the shoreline. However, in some areas of the world, as the front lines of swells arrive at the shallow waters with different angles, they tend to turn from deep water towards shallow water. This phenomenon occurs because the shallow water depth actually reduces the wave travelling speed, while the other part of waves moving in deep water continue to move at the same speed.

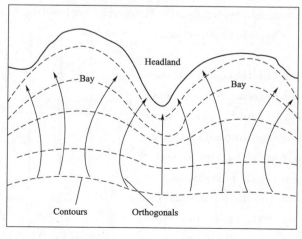

Figure 3 Wave refraction

Chapter 3 Coastal Engineering

Wave reflection

Reflection may occur when waves propagate to solid obstacles, such as breakwaters, seawalls, cliffs or sloping beaches. For vertical and rigid structures, the fraction of reflected wave energy can be large. Reflection is much less for permeable structures or mild slopes. On the seaside of a reflected coast, the wave motion is a superposition of the incident wave and the reflected wave. A well-known result of such superposition is the generation of a standing wave, which can be found when a monochromatic wave train with normal incidence reflects perfectly from a vertical seawall. Very often, the reflection of coast is less than 100%, thus it produces a partial standing wave rather than a perfect standing wave.

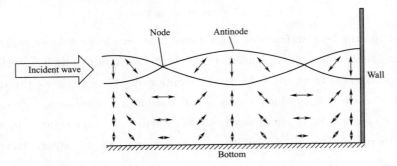

Figure 4 Standing wave due to wave reflection

Wave diffraction

When a wave encounters an obstacle during the propagation, it can change direction or wrap around it. In oceans, this can happen when a wave meets an object like a jetty and the wave rotates around it (sometimes diffraction occurs when the wave passes through a small gap in a breakwater or moves between two islands). For waves with longer wave length (i.e., longer period), the "wrapping" or turning effect of a wave is larger. Diffraction can occur in both shallow or deep waters and it is different from refraction because it is not caused by the change in water depth. However, both refraction and diffraction lead to a change in the wave travelling direction.

Wave breaking

Waves approaching to the coast steepen as the water depth decreases. As the wave steepness reaches a threshold value, wave breaking occurs with a large amount of energy dissipated, causing a rise of the mean water level (known as wave setup). The surf zone is a region extending from the incipient breaking point to the limit of wave uprush. Wave breaking can be classified into three main types: spilling breaker, plunging breaker and

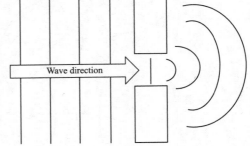

Figure 5 Wave diffraction

surging breaker.

Spilling breaker

When a wave passes through a gently sloping seabed close to the beach, spilling breaker will occur. The wave breaks slowly and over a long distance. White water spills from the wave crest and drops down to the wave front, losing its energy.

Plunging breaker

When a wave approaches to a moderate bottom, plunging breaker will occur. The wave becomes steeper than the spilling breaking wave, and the crest forms a well-defined curl and falls forward with considerable energy. When this kind of wave strikes the coastline at an angle and propagates across the coastline, the "tube" created by the curl of water mass becomes a favorite among surfers.

Surging breaker

When a long-period, small-amplitude wave propagates to a steep beach, surging breaker will occur. The wave crest does not spill or curl; it accumulates and then collapses rapidly onto the beach with less foam or spray compared with other two breakers.

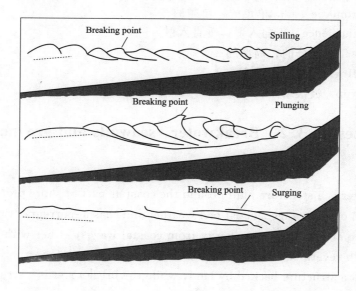

Figure 6　Main types of breaking waves

New words and expressions

1. shear　n. 剪切
2. resonance　n. 共振
3. propagate　v. 传播
4. propagation　n. 传播
5. swell　n. 涌浪
6. wind seas　风浪

7. dispersion   n. 分散，色散
8. wave transformation   波浪变形
9. attenuation   n. 衰减
10. wave shoaling   波浪浅化
11. unaltered   adj. 不变的
12. transitional   adj. 变迁的，过渡的
13. dissipation   n. 耗散
14. oscillatory   adj. 摆动的，震荡的
15. gradient   n. 梯度
16. equivalent   adj. 相等的，等价的
17. wave refraction   波浪折射
18. permeable   adj. 可渗透的
19. superposition   n. 叠加，重合
20. incident wave   入射波
21. standing wave   驻波
22. monochromatic   adj. 单色的，单频的
23. normal incidence   正向入射，垂直入射
24. partial   adj. 部分的，局部的
25. wave diffraction   波浪绕射
26. wave setup   波浪增水

Lesson 3.2   Breakwater, seawalls and groynes

Breakwater

A breakwater is a structure built around the coast as part of shore protection to defend the harbor basin or anchorage against bad weather and coastal currents. Breakwater structures are built to absorb the wave energy from coastal waves, either with blocks (such as caissons) or with revetment slopes (such as rocks or concrete armor units). In general, the revetment is a structure backed by land, while the breakwater is a structure backed by water, i.e., there is sea on both sides of the structure.

Rubble mound breakwater

To dissipate wave energy, rubble mound breakwaters take advantage of the structure's voids. A rubble mound breakwater consists of a pile of stones grouped by the unit weight: small stones function as the core, while large stones serve as a protective layer to protect the core against wave impact. Armor units such as rocks or concrete blocks outside the structure absorb a large proportion of wave energy, whilst gravels and sands impede the rest of the energy from going through the core of breakwater. Depending on the used materials, the slope of a revetment normally ranges from 1 : 1 to 1 : 2. In shallow waters, the cost of building a revetment breakwater is usually cheaper. As the water depth deepens, the required materials thus the cost increases significantly.

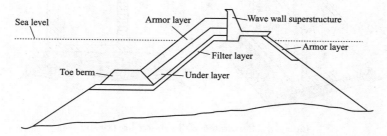

Figure 1 Rubble mound breakwater

Caisson breakwater

Caisson breakwaters generally have vertical side walls and are constructed where ships and other water vessels need to be berthed. The caisson size as well as its mass are designed to resist the overturning force of the incident waves. It is relatively expensive to build caisson breakwaters in shallow waters, but the construction of caisson breakwater can save a lot of money at deeper locations. To dissipate wave energy, hence reducing wave reflection and dynamic pressure on its vertical wall, a rubble mound is sometimes added in front of the caisson. This treatment offers more protection on the seaside wall of the breakwater, but it also facilitates wave overtopping over the structure.

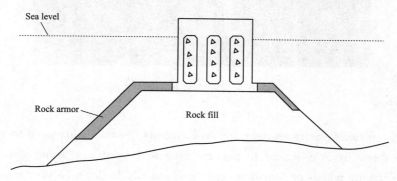

Figure 2 Caisson breakwater

Perforated-wall caisson breakwater

The wave-absorbing caisson is an identical but more advanced design. Different kinds of perforations are opened in its seaside wall. This structure has been successfully applied in offshore oil exploration as well as coastal engineering using low-crested structures. The primary advantages of perforated-wall caisson breakwaters are the saving of construction cost in deeper waters and less blockage to surrounding water environment.

Floating breakwater

Compared with the traditional fixed breakwater, the floating breakwater is an alterna-

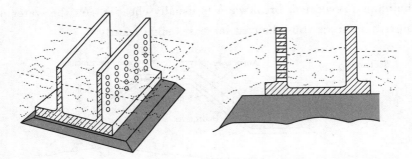

Figure 3   Perforated-wall caisson breakwater

tive to protect a coast from wave action. In coastal areas with mild wave conditions, such method is effective. Therefore, they have been increasingly used to protect small ports or docks, and less frequently, to reduce coastal erosion. Floating breakwater is generally divided into four categories: box, pontoon, mat and tethered float.

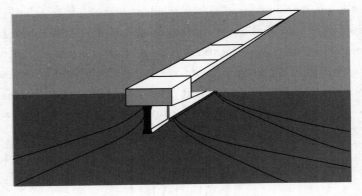

Figure 4   Floating breakwater

Seawall

In reality, seawall and revetment are synonymous. Seawalls are used to separate land areas from waters. It is designed to prevent coastal erosion and other damages (e.g., flood) from extreme waves or storm surges, it is usually built to be very massive.

Seawalls are constructed along the shoreline or at the foot of cliffs or dunes. A seawall is generally a sloping concrete structure. Its seaside face can be smooth, stepped or curved. It can also be built as a rubble mound structure or as a concrete, steel or wood structure. One characteristic of seawall design is that the structure can withstand strong wave action and storm surge. There is a great similarity between rubble mound seawall and rubble mound revetment, which is usually used to protect the foot of non-flexible seawall. However, a revetment is generally used as a supplement to seawalls or as a separate structure at less wave exposed shores. Sometimes, an exposed dike with reinforcement to resist wave impact can also be treated as a seawall.

Lesson 3.2 Breakwater, seawalls and groynes

The seawall is a passive protective structure that prevents the coast from erosion and flooding. Seawalls are frequently used for the wave-affected coastal zones in cities, where land areas are scarce and require good protection, and promenades are usually built at the top of these structures. They are also applied in some less populated coastal areas, especially in areas where there is an urgent need for a combination of coastal and marine defense.

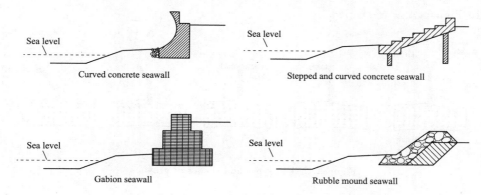

Figure 5 Main types of seawalls

Groyne

A groyne is a coastal structure stretching from the coast to the seawater. It is generally built normal or somewhat inclined to the coastline. For the purpose of shore protection, it reduces coastal erosion due to oblique waves and longshore currents by promoting sediment deposition.

The design of a groyne (planform, length, height, cross-shore profile, and inclination angle, etc.) has a prominent impact on the coastal morphology, which is also related to sea level, wave condition and surf-zone sediment transport. Shore protection using only one groyne is very inefficient. Therefore, a typical shore protection project usually uses several to dozens of individual groyne structures to form a group.

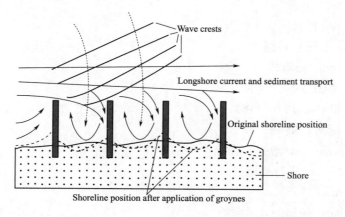

Figure 6 Interactions among groynes, waves, longshore currents, and shore

Chapter 3　Coastal Engineering

The height of a groyne affects its retained sediments due to longshore transport. A given groyne can be effective in either emerged or submerged condition, depending on the sea level change caused by tide or storm surge. Typically, a groyne is designed to be 0.5 to 1.0 m higher than the mean sea level. If it is too high, the increased reflected waves may lead to local scouring. As for the planform, a groyne can be straight, curved, L-shaped, T-shaped or Y-shaped.

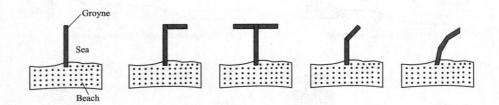

Figure 7　Main types and shapes of groynes

New words and expressions

1. anchorage　n. 锚地
2. armor　n. 装甲，防护
3. rubble mound breakwater　抛石防波堤
4. berm　n. 坡台，滩台
5. void　n. 空隙
6. caisson breakwater　沉箱防波堤
7. overturning force　倾覆力
8. wave overtopping　越浪
9. perforated-wall caisson breakwater　开孔墙式沉箱防波堤
10. perforation　n. 穿孔，打孔
11. blockage　n. 阻塞
12. pontoon　n. 浮筒
13. tether　v. 系链，拴绳
14. extreme　adj. 极端的
15. dune　n. 沙丘
16. supplement　n. 增补（物）
17. dike　n. 堤防，堤坝
18. reinforcement　n. 增强，强化
19. promenade　n. 人行步道
20. groyne　n. 丁坝
21. oblique　adj. 倾斜的，斜向的
22. scouring　n. 冲刷

Lesson 3.3 Coastal morphology

There are two main types of coastal morphology: one is erosion and the other is deposition. They exhibit significantly different landforms, although each type may contain some features of the other. Generally, the erosional coast means that there is little or no sediment, while the depositional coast is characterized by a long-term accumulation of abundant sediment.

Landform types of erosional coasts

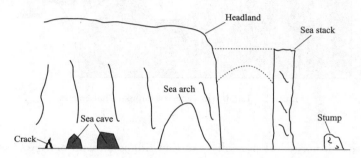

Figure 1 Landform types of erosional coasts

Headland
A headland is a coastal landform, usually a place where the land is high and extends into the water through a steep descent. A headland with a fairly large size is usually known as a cape. Headlands are often featured by high terrain, steep cliffs, rocky coasts, breaking waves, and strong erosion.

Sea cave
A sea cave is a type of landform mainly shaped by wave impact from the ocean. The most relevant cause is erosion. Sea caves have been discovered worldwide. They form along the current shorelines, leaving residual eroded caves on the previous shorelines. Some of the world's largest erosive caves due to waves are found on the Norwegian coasts, but they are 100 ft or higher than the present sea level.

Sea arch
Another spectacular erosional morphology is the sea arch, which is formed by erosion at different rates, typically due to different resistance of the bedrocks. These arches may be in arcuate or rectangle shape with the openings extending below the water level. The height of an arch can be tens of meters above the sea level.

Sea stack
Sea stacks are geological landforms composed of steep or even vertical rock columns located in the sea close to the coastline. It is formed via wave erosion, some are as high as several meters and form isolated pinnacles on the relatively smooth face of wave action.

Since erosion is a continuous process, these features are not permanent. Erosion will eventually cause the stack to collapse, leaving a stump.

Landform types of depositional coasts

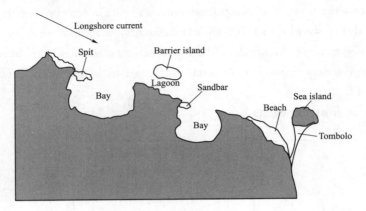

Figure 2   Landform types of depositional coasts

Sandbar

Sandbars are submerged or partially exposed ridges of sand or coarse sediments transported from the beach to the sea by waves. The turbulence generated from the breaking waves on the beach can scour a trench on the sandy seabed, some sand is transported forward to the shoreline, and the remaining is deposited on the offshore side of the trench. The suspended sand in both backwash and rip currents also contributes to the formation of the bar, as does some moving sand from the deeper water to the coast. Because of the strike of breaking waves on the sandbar, its top is usually maintained below the still water.

Tombolo

Far from the river mouths, deposition also occurs in areas where wave exposure is weaker. For instance, a small island near the coast can keep the mainland coastline free from large waves, hence reducing wave action between the island and the mainland. Such an area is known as a wave shadow zone. A longshore current is unable to pass through such a calm area in that the movement of waves is required to drive the longshore current. Thus most sediments carried by the longshore current into this region are deposited here. Sedimentary deposits finally form a sand bridge between the island and the mainland. Such kind of bridge is called a tombolo.

Spit

The sediments carried by the longshore current also cause an extension of sandy beaches, known as spit. Spit can extend partially to the entrance of a cove or bay. Once the longshore current meets the cove or bay, a change of drift path directs the longshore current to the deeper water. However, with the increase of water depth, wave motion near the sea bottom declines, hence the drift cannot be kept. Then the carried sediments are deposited to form a spit. As soon as the spit is formed, wave impact on the seaside of spit is still very strong, and the longshore current keeps on carrying sediments to the spit end where they are settled down. The spit defends a bay against the waves, providing a sheltered region of quiet water on its shoreward side. Accretion at the end of a spit in the shadow zone may drive the spit to shift landward, forming a hook. Furthermore, a loop appears when the spit further develops and ends at the shore.

Barrier island

Barrier islands can also protect the mainland against the waves. They are long but narrow islands parallel with the coastline of the mainland. A shallow bay, frequently known as a lagoon, usually exists to separate the barrier island from the mainland. The barrier island possibly results from a well-formed spit or a sand dune later submerged by the sea level rise. Owing to the effect of wind and/or water, the barrier island is usually in a migratory state.

New words and expressions

1. morphology   n. 形态（学）
2. landform   n. 地形，地貌
3. headland   n. 岬角
4. cape   n. 海角
5. terrain   n. 地形，地势
6. sea cave   海蚀洞
7. sea arch   海蚀拱
8. sea stack   海蚀柱
9. pinnacle   n. 高峰，尖峰
10. stump   n. 残余部分
11. sandbar   n. 沙洲，沙坝
12. ridge   n. 脊，突起
13. transport   v. 运送，传输
14. scour   v. 冲刷
15. trench   n. 沟槽
16. rip current   裂流，离岸流
17. tombolo   n. 连岛沙洲
18. longshore   adj. 沿岸的
19. spit   n. 沙嘴
20. cove   n. 小海湾，凹岸
21. accretion   n. 堆积，淤积
22. hook   n. 钩状（物）
23. loop   n. 环状（物）

24. lagoon   n. 潟湖
25. migratory   adj. 迁移的

Lesson 3.4 Coastal sediment transport and beach nourishment

Coastal sediment transport

The interactions between different coastal landforms and complicated physical processes cause the transport of coastal sediments, which is also called littoral transport. It is usually decomposed into longshore transport and cross-shore transport.

Longshore sediment transport

Coastal sediment transport is largely dependent on the longshore current generated by oblique waves breaking along the coastline. Longshore drift is caused by the longshore current. This geological process includes the transport of sediments (clay, silt, pebble, sand, etc.) along the shore parallel to a coastline, depending on the direction of the oblique incoming wind. Such wind blows the seawater along the shore, generating a current moving parallel to the shore. Consequently, a movement of sediment which is also known as longshore drift is driven by the longshore current. Such a process mostly happens within the surf zone.

In those oblique windy days, the beach sand is also mobilized because of seawater swash and backwash on the beach. Breaking waves carry seawater uprush (swash) with an inclined angle whereas gravity causes it to fall (backwash) perpendicular to the coastline. Therefore, sand on the beach can move in a zigzag way towards the direction away from the beach, dozens of meters a day. Such a phenomenon is called beach drift. However, some people think it is simply a form of longshore drift in that the total sand transport is parallel to the shore.

Longshore drift determines the size of many sediment particles because it acts in a little different manner, relying on the nature of the sediment, for example, the difference in the longshore drift of sediments between a sandy beach and a shingle beach. The sediment is greatly influenced by the oscillatory behavior of nearshore waves, because the sediment movement can be caused by the bed shear from breaking waves and the longshore current. Because sandy beach is much gentler than shingle beach, plunging breaker is more likely to occur on the latter. Owing to the absence of an extended surf zone, most longshore transport happens in the swash zone.

Cross-shore sediment transport

Longshore sediment transport is mainly caused by the wave-generated longshore current, while cross-shore transport is caused by the seawater movement associated with waves and undertow. Seasonal shoreline change associated with seaward sediment trans-

Lesson 3.4 Coastal sediment transport and beach nourishment

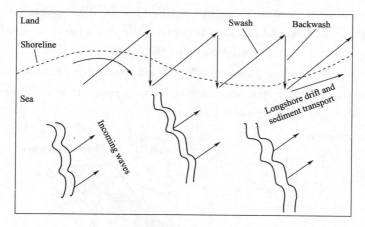

Figure 1 Longshore sediment transport

port and sediment deposition near the offshore sandbar are often considered as a response to winter storms. Scholars have no consensus on the accurate causes of the sandbars. But it is in general believed to be the result of the undertow due to breaking waves. As wave height resulted from the storm surge increases, a sandbar forms farther away from the shore. As the water depth becomes larger, the overall scale of the sandbar also increases, requiring a larger amount of sand. Part of the sand is offered by eroding the subaerial part of the beach profile.

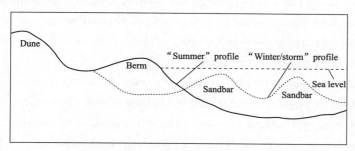

Figure 2 Cross-shore sediment transport

Beach nourishment

Beach nourishment is an adaptive technology mainly used to cope with shoreline erosion, and such technology may also be beneficial to reduce coastal flooding. This is a soft engineering approach for coastal protection by artificially adding sediment with appropriate quantity to the beach area where there is a sand deficit. Sometimes, beach nourishment is also known as beach fill, replenishment, renourishment, and beach feeding.

In the process of beach nourishment, there are two main physical processes that affect its performance (that is, the effective increase of beach width during the nourishment), each of which runs on different time scales. The first physical process, the artificially altered beach profile during artificial nourishment, will re-establish an equilibrium profile by the offshore sediment transport on a newly filled beach. This process occurs in a period of months to years. The second physical process is the disturbance of the beach shoreline artificially altered by the beach nourishment to the longshore current, which leads to the sand being transported to other places near the filled beach during a long and slow coastal sediment transport process. For a beach nourishment project that lasts long enough, this process occurs on a time scale of years to decades.

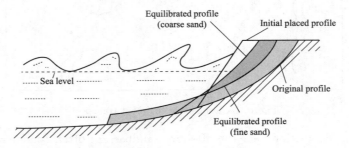

Figure 3　Physical processes involved in beach nourishment

　　One principle in the design of a beach nourishment project is to conform to the dynamic process of a natural beach, so that the filled sand can move continuously and respond to varying waves and water levels. Depending on local conditions, coastal engineers may choose to place filled sand directly on the beach, as underwater mounds, as dunes or all three. The sand, once filled, is redistributed with the coastal dynamic process and then redistributed near the coast, leading to a change of the shoreline response. It ultimately achieves the goal of increasing the beach width, adjusting the beach slope and improving the shoreline protection provided by the beach.

　　The slope of the beach treated by the nourishment project becomes gentle, and more energy is dissipated through wave breaking in shallow water when waves propagate shoreward, thus wave energy reaching the shoreline is reduced. The coexistence of undertow and longshore current drives the sediment transport in both offshore and longshore directions, carrying some newly filled sand to the deeper waters or form the longshore sediment transport. These sentiments carried by offshore and longshore transports often form an offshore sandbar that causes waves to break farther offshore, enhancing the dissipation of wave energy reaching the shoreline.

　　In order to ensure that a nourished beach can continuously provide sufficient beach width and shoreline protection in extreme sea conditions caused by occasional storms, beach nourishment must be carried out periodically. Continuous supplement of new sand to the beach for a long-term sediment loss is called periodic beach renourishment.

New words and expressions

1. beach nourishment　海滩养护
2. littoral　adj. 近岸的，沿海的

Lesson 3.4 Coastal sediment transport and beach nourishment

3. drift   n. 漂流，漂移
4. clay   n. 泥土，黏土
5. silt   n. 淤泥，粉砂
6. pebble   n. 卵石
7. surf zone   破波带
8. swash   n. 上冲流
9. backwash   n. 回流，下泄流
10. inclined   adj. 倾斜的
11. zigzag   adj. 曲折的，之字形的
12. particle   n. 颗粒
13. shingle beach   砾石岸滩
14. swash zone   冲流带
15. cross-shore   adj. 向岸的
16. undertow   n. 海底回流
17. offshore   adj. 离岸的，近海的
18. subaerial   adj. 近地面的，陆上的
19. deficit   n. 亏损，不足
20. replenishment   n. 补充，补给
21. mound   n. 堆状物，堆积

Chapter 4　Ocean Engineering

Lesson 4.1　Offshore platforms

　　Offshore platforms, also called oil platforms, are giant structures that can drill well and exploit oil and gas in the deep ocean. These platforms have specialized facilities to store crude oil and gas until they are shipped to a refinery. Some have corresponding facilities for the labor force accommodation as well. Based on the requirements, offshore platforms can be floating or fixed to the seabed.

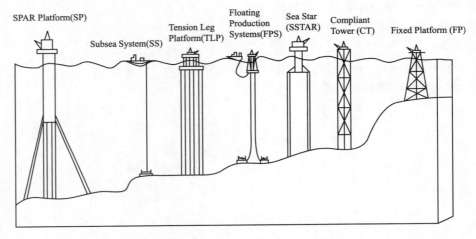

Figure 1　Main types of offshore platforms

Main types of offshore platforms
Fixed platform
　　Fixed platforms are constructed with huge steel or concrete supporting legs mounted directly on a seabed. The fixed platform has enough space to place drilling rigs as well as production equipments, and it can also provide accommodation for crew members. Such kind of platform is highly stable and built for long-term service. Usually, it can be established in waters of 520m (1700ft) deep. When the water depth is beyond this range, the high cost causes the platform to lose its practical value.
Gravity-based structure (GBS)
　　A GBS is a steel or concrete structure, normally installed directly to a seabed. Steel GBS is mainly applied in areas without crane barges, or with crane barges that are not

suitable for the installation of traditional fixed offshore platforms, such as in the Caspian Sea. Examples of existing steel GBS all over the world can be seen in offshore Turkmenistan waters (Caspian Sea) and offshore New Zealand. Steel GBS generally has no capacity to store hydrocarbons. GBS is installed by pulling it out of the dock using wet-tow or/and dry-tow. It is then self-installed via controlling the ballast of its compartments with seawater. To locate the GBS when it is installed, it can be linked to a transportation barge or another kind of barge by a strand jack (as long as the size of the barge is sufficient to support the GBS). When loading the GBS, the jack should be released slowly to guarantee that the GBS only sways a little bit from the designed position.

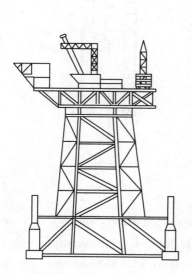 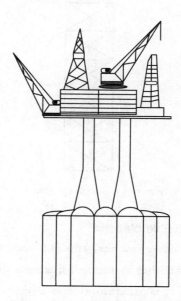

Figure 2　Fixed platform　　　　Figure 3　Gravity-based structure

Compliant tower

　　Compliant tower is built in view of the design idea for a fixed platform. However, it employs a narrow tower structure made of concrete and steel. Such a platform is designed to be flexible and can swing or shift laterally with both wind and wave forces. The compliant tower is able to operate in waters varying between 457m to 914m (1500ft to 3000ft) deep.

Spar platform

　　In the design of this platform type, the platform is installed on a large cylindrical hull which is hollow inside. The bottom of the cylinder can reach a water depth of around 213m (700ft). Although the cylinder structure ends far away from the seabed, the platform atop it can still stay in place because of the mass of the cylinder itself. This type of platform can operate at a depth of 3048m (10000ft) in water.

Chapter 4　Ocean Engineering

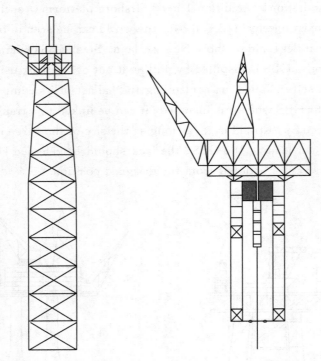

Figure 4　Compliant tower　　Figure 5　Spar platform

Semi-submersible platform

As the name implies, this is a semi-submerged platform, and can be moved from one location to another as needed. Because the lower hull of the semi-submersible platform is submerged, it is, therefore, less affected by the wave load than an ordinary ship. However, because of the small water-plane area of the semi-submersible platform, it's very sensitive to load changes. Therefore, stability must be maintained by adjusting its onboard mass. Unlike fully submerged structures, the semi-submersible platform is not supported on the seabed. It relies on the principle of dynamic positioning and uses large anchors to keep it in position. This type of platform can work in a water depth with a range of 60m to 3000m (200ft to 10000ft). HAI YANG SHI YOU 981 is the 6th generation deep-water semi-submersible platform and the first drilling rig using the 3rd generation dynamic positioning system (DP3) in China. Its designed displacement is 30670t and its average draught is 19m. Its overall length is 114.07m and its width is 78.68m.

Jack-up drilling rig

A jack-up drilling unit (also called jack-up), as its name implies, is a device that can lift the rig up by lowering the leg, just like a jack. Although some designs allow the device to work at a depth of 170m (560ft), such devices are generally used in waters above 120m (390ft) deep. The jack-up drilling rig is able to move from one position to another, and can subsequently be fixed through a pinion-and-rack gear system on each leg to deploy the leg to the bottom of the sea.

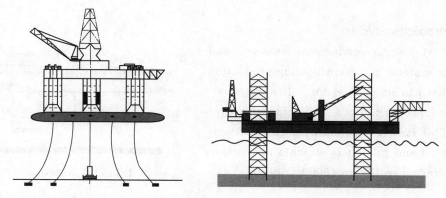

Figure 6　Semi-submersible platform　　　Figure 7　Jack-up drilling rig

Sea star platform

This kind of platform is designed as a larger version of semi-submersible platform. However, it is connected to the seabed through flexible steel legs instead of anchors. Such platforms typically operate in a water depth covering from 152m to 1067m (500ft to 3500ft).

Tension leg platform

A tension leg platform falls into the category of floating platforms. In addition to its unique tension leg extending from the seabed to the platform itself, it can usually be regarded as a giant version of the sea star platform. Such kind of platform can operate in the water up to a depth of 2134m (7000ft).

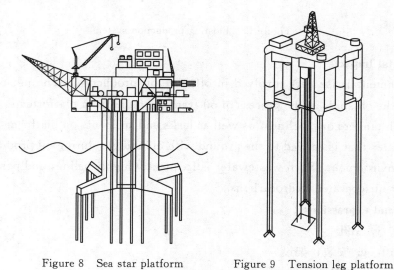

Figure 8　Sea star platform　　　Figure 9　Tension leg platform

Drillship

A drillship is a marine ship equipped with drilling facilities as well as dynamic positioning systems, and it can maintain its position above the oil well. This type of ship is generally used for exploration and drilling. It can operate in a water depth up to 3700m (12000ft).

Floating production system

FPSO (floating production, storage, and offloading system) is a main floating production system that can be applied for a floating semi-submerged platform or a drillship when required. This system is primarily used to process and store oil and gas. It is suitable to work in the seawater up to 1829m (6000ft) deep.

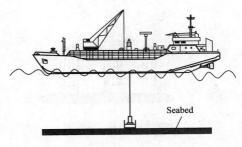

Figure 10　Drillship

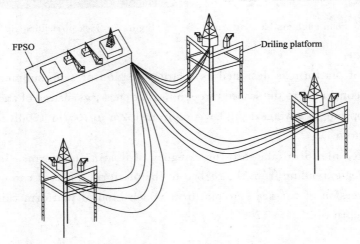

Figure 11　Floating production system

Environmental impacts

Environmental risks are involved in offshore oil production. The most notable risks come from the oil spills in the process of oil transportation from platforms to onshore facilities through tankers or pipelines, as well as leaks and accidents on platforms. In addition, the wastewater that is carried to the ground with oil and gas during oil production can also affect the environment. Such wastewater is frequently highly saline, and possibly contains dissolved or unseparated hydrocarbons.

New words and expressions

1. drill　v. 钻孔
2. mount　v. 安装，架设
3. drilling rig　钻机，钻探装置
4. ballast　n. 压舱（物）
5. compartment　n. 隔间，舱室

6. strand jack   绞索千斤顶
7. compliant tower   顺应塔式平台
8. hull   n. 船体
9. semi-submersible   adj. 半潜式的
10. water-plane   水线面，水平面，吃水面
11. onboard   adj. 船上的，船载的
12. anchor   n. 锚
13. jack-up drilling rig   自升式钻井平台
14. pinion-and-rack   齿轮齿条
15. tension leg   张力腿
16. drillship   n. 钻井船
17. spill   n. 溢出，泄漏
18. saline   adj. 含盐的

Lesson 4.2 Wave energy

Marine energy includes the energy associated with ocean waves, tides, salinity, and temperature differences. The seawater motions in the ocean all over the world contain a large amount of kinetic energy. A portion of this energy can be used to generate electricity for global households, transportation, and industry.

Wave energy, also called ocean wave energy, is a kind of renewable energy. It uses wave power to generate electricity. In contrast to tidal energy, which utilizes both flood and ebb tidal currents, wave energy generates electricity by using the kinetic energy from periodic horizontal and vertical movements of the sea surface associated with the action of waves. The operation principle of most wave energy devices is to convert the up-and-down motion of ocean waves into mechanical energy that can be used by the power generation device (usually referred to as a power take-off device, or PTO for short), and then drive the device to generate electrical power. Table 1 shows the global distribution of wave energy.

Table 1   The distribution of wave energy resources worldwide

Wave Power Flux, kW/m	Country, Region	Sea, Ocean
1	Colombia	Caribbean Sea
1.1	Turkey	Black Sea
2	Iran	Persian Gulf
4.6	Lebanon	Mediterranean Sea
5.1	Shandong Peninsula (China)	Yellow Sea
5-14	Iran	Caspian Sea

Continued table

Wave Power Flux, kW/m	Country, Region	Sea, Ocean
15 – 30	Hawaii	Pacific Ocean
20 – 40	Norway	Atlantic Ocean
25	States of Oregon and Washington (USA)	Pacific Ocean
25	Southeast Africa	Indian Ocean
25 – 30	Canary Islands	Atlantic Ocean
30	Bay of Biscay	Atlantic Ocean
30 – 50	Western Australia	Indian Ocean

So how does it work? The following diagram shows the principle of wave energy and how to use it to generate electricity.

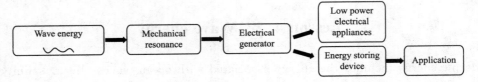

Figure 1　The principle of wave energy to generate electricity

Based on the above wave energy conversion principle, there are several typical wave energy generation technologies (according to the definition by the European Marine Energy Centre, EMEC), the most common ones are described as follows.

Attenuator

The attenuator is a floating device arranged along the direction of wave propagation. The technical principle of this device allows it to extract wave energy from the incident waves with lower power. The floating tube section of the device moves with the motion of waves, thus driving the hydraulic system used for energy conversion to generate electricity. All mechanical equipment for power generation is placed inside the watertight tube section. This device is usually made up of two or more tube sections connected to each other and can be deployed in various water depths. The more tube sections connected to each other, the greater power generation it achieves.

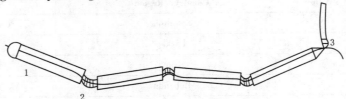

Figure 2　An example of the attenuator
1—tube; 2—hinged joint; 3—cable

Point absorber

The point absorber is also a floating device, but its configuration allows energy to be obtained from incident waves in all directions. In addition, the device is not very large and it has a shape like a typical point buoy, thus allowing a few similar devices to form an array to extract energy from the same wave. The buoy can convert the incident wave energy into the kinetic energy of the oscillator, hydraulic mechanical energy, or pneumatic energy to propel the generator. The PTO system can be installed in buoys or arranged to the bottom of the sea. In order to maintain the position of the device, the point absorber is generally anchored to the sea bottom, from where its cable is extended onshore. The power generation efficiency of the point absorber is generally between 30% and 45%.

Oscillating wave surge converter

The oscillating wave surge energy converter utilizes the surging wave motion to produce electricity. Such devices usually have a permeable or impermeable floating interceptor to capture the wave energy. The floating interceptor is normally hinged on the base sitting on the seabed. The kinetic energy carried by waves is intercepted by the interceptor, so that the interceptor rotates around its hinge in the water to form an oscillation. The interceptor is usually connected to the base by a hydraulic, pneumatic, or linear mechanical device as the PTO, thus converting the kinetic energy of the interceptor into electric power. Considering that the interceptor of such devices must be able to intercept wave energy near the sea surface, such devices can generally only be deployed in shallow waters, otherwise, the interceptor will be too large, reducing the reliability and waters efficiency of the device.

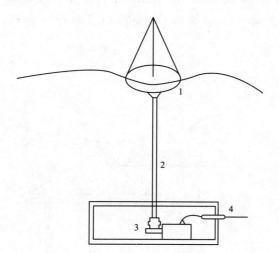

Figure 3 An example of the point absorber
1—buoy; 2—PTO; 3—driver system and generator; 4—cable

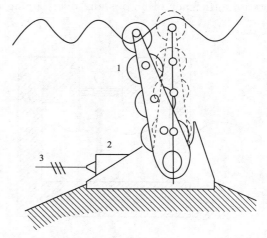

Figure 4 An example of the oscillating wave surge converter
1—buoy; 2—PTO and generator; 3—cable

Chapter 4　Ocean Engineering

Oscillating water column

　　This device employs air circulation to drive the generator. Wave energy devices with this technology can be placed both onshore (easy to achieve large installed capacity) and offshore as floating devices (relatively smaller installed capacity). Such kind of device usually has a closed air chamber, the lower part of which is connected to the sea, and the upper part is airtight except for the orifice connected to the PTO. The motion of waves drives the water surface in the chamber to move up and down. The oscillating water surface acts like a "weightless piston", which creates pressure oscillations in the chamber by compressing and expanding the chamber volume, thus driving air through the only opening in the upper part of the chamber. The PTO near the opening usually uses a Wells turbine or impulse turbine, which converts airflow from the chamber or from the outside to a unidirectional torque on the turbine shaft thus reducing the energy loss.

Overtopping device

　　The overtopping wave energy device captures the overflow produced by the incident wave runup over a slope. The overtopped waters flow into a reservoir that is higher than the sea surface, thus converting the kinetic energy associated with waves into the gravitational potential energy. Under gravity, these water masses then go down and eventually flow back into the sea through a PTO (usually a hydro-turbine) installed at the reservoir bottom. In order to improve the efficiency of such kind of device, wave energy focusing can be achieved by designing an auxiliary structure in the shape of a horn. The overtopping device can be installed onshore or in the sea as a floating device. Normally, because of the wave energy dissipation associated with wave breaking near the shoreline, the power generation efficiency of an onshore overtopping device is less than that of an offshore device.

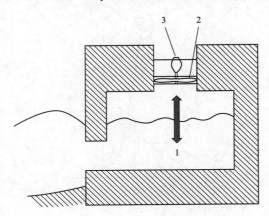

Figure 5　An example of the oscillating water column
1—air chamber; 2—air turbine; 3—generator

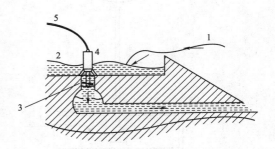

Figure 6　An example of the onshore overtopping device
1—sea; 2—reservoir; 3—hydro-turbine; 4—generator; 5—cable

Lesson 4.3 Submarine pipelines

New words and expressions
1. salinity   n. 盐度
2. kinetic energy   动能
3. renewable   adj. 可再生的
4. mechanical energy   机械能
5. take-off   获取
6. attenuator   n. 衰减器
7. tube   n. 管子，管状物
8. watertight   adj. 水密的，防水的
9. absorber   n. 吸收器
10. buoy   n. 浮标
11. oscillator   n. 振荡器
12. pneumatic energy   气动能
13. wave surge   浪涌
14. interceptor   n. 拦截器
15. orifice   n. 孔口
16. turbine   n. 涡轮
17. unidirectional   adj. 单向的
18. torque   n. 扭矩
19. shaft   n. 传动轴
20. overtop   v. 溢出，漫顶
21. overflow   n. 溢流
22. wave runup   波浪爬高
23. auxiliary   adj. 辅助的

Lesson 4.3   Submarine pipelines

A submarine pipeline, also called a subsea, marine, or offshore pipeline, is usually laid on the seabed or in the trench beneath the bed. Sometimes, the pipeline is primarily onshore. While in some other sites, it passes through small water areas such as bays, straits, and rivers. Submarine pipelines are mainly employed to transport oil or gas, and water delivery is equally important.

Route selection

The first and most important task in the planning of a submarine pipeline is to select the route. Such a selection must consider various issues, some of which are political, but others are primarily related to geological hazards, physical factors on the way as well as other seabed utilization in the considered region.

Physical factors

The main physical factor in constructing a submarine pipeline is the seabed condition, i. e., whether the seabed is smooth (it is comparatively flat) or rough (it is uneven and has points with different elevations). For an uneven seabed, the pipeline has a free span that is not supported when connecting two high points. If the unsupported section of a pipeline is too long, the bending stress acting on it becomes too strong as a result of its weight. Vibration resulted from vortexes associated with water flow can cause a problem as well. Methods for a correction to the unsupported pipe span involve the bed leveling as well as the post-installation reinforcement, e. g., berm building or sand filling beneath the pipe. The bed strength is also an important factor. When the soil is not sufficiently hard, the pipe may subside to some degree, causing inspection, maintenance, and preset tie-ins hard to be performed. In another extreme case, the cost of trenching the rock seabed is high, and the abrasion of pipeline external coating and subsequent damage can occur at a high point. An ideal situation is that the soil allows the pipeline to settle in it to a certain extent, thus providing some lateral stability for the pipeline.

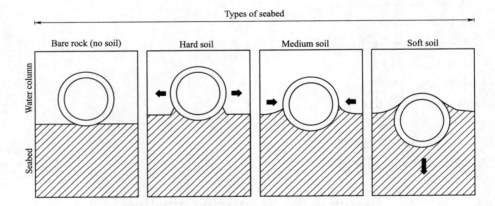

Figure 1 A submarine pipeline interacting with the seabed

The rest of physical factors that should be considered before pipeline construction are:

(1) Seabed mobility: seabed features such as sand waves and large sand ripples move over time, thus a pipeline supported by the top of these features during construction may be found in a trench when the pipeline is in a later operation. The development of those features is hard to estimate, so it is better to avoid pipeline building in their existing areas.

(2) Submarine landslides: they are caused by higher deposition rates occurring on steep slopes, and can be initiated by earthquakes as well. As the soil around a pipeline slips, especially when the induced shift is at a large angle to the pipeline, pipes in the soil may be subjected to serious bending and tensile failure.

(3) Currents: strong current is annoying because it hinders the laying of pipelines.

For example, in shallow seas, the tidal current between two islands can be quite strong. In this case, it is better to set the pipeline in other places, even though the new route is longer.

(4) Waves: waves (in extreme wave conditions) are also very troublesome for pipeline laying in shallow waters. Because of the scouring effect from the wave-induced flow, it is also unfavorable to the stability of the pipeline in operation. This emphasizes the importance of pipeline landing (where the pipeline arrives at a coastline) which requires a careful choice of the location.

(5) Ices: in freezing seawaters, ice floats usually drift into shallow waters with their keels touching the sea bottom. When drifting continuously, they may erode the seabed then strike the pipe. The icicle may also destroy the pipe via imposing large local stress on it or inducing the failure of soil around it. Ice hole is an additional risk for the pipeline in cold waters. The seawater ejected from it can carry away the soil beneath the pipe, causing it susceptible to excessive stress as a result of the self-weight or vibration associated with the vortexes. In regions where the risks are identified, the pipeline can be designed to lie in a backfilled trench during the route planning.

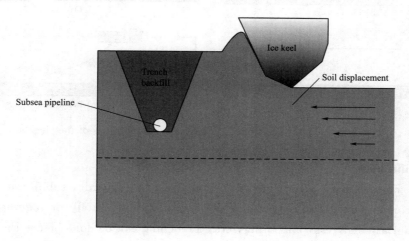

Figure 2   Interaction between a submarine pipeline and an iceberg

Pipeline construction

Pipeline construction includes two steps: connecting a lot of pipe sections to a full line, then laying the line along the designed path. There are several ways to lay the submarine pipelines. They are chosen according to the following principles: physical environment (e. g. , waves, currents), facility applicability and cost, seawater depth, pipeline length and pipe diameter, limitations associated with other pipelines and structures already existing on the way. Current laying approaches are normally classified into four types: pull/tow, S-lay, J-lay, and reel-lay.

The pull/tow system

For this system, a submarine pipeline is installed ashore and subsequently towed to the specified position. The assembly can be parallel or normal to a coastline. As for the former, the entire pipeline can be connected before towing and laying. For towing procedures, the following approaches can be used: surface tow, near-surface tow, mid-depth tow, and bottom tow.

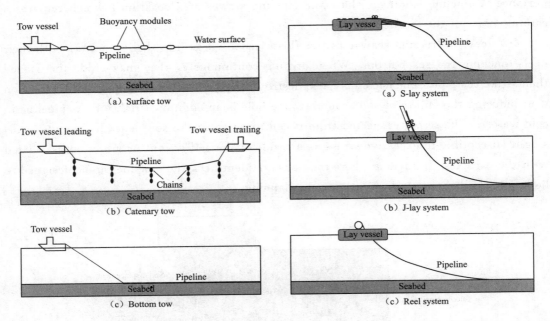

Figure 3　Types of the pull/tow systems　　　Figure 4　Other types of the pipeline laying systems

Pipeline stabilization

Several methods have been adopted to stabilize or safeguard a submarine pipeline as well as its components. Those methods can be utilized separately or together, such as trenching, burial, mattress, anchoring, etc. Trenching can be done before the pipeline is laid (pre-laying trenching) or by removing soil beneath the pipe (post-laying trenching). As for the latter, a trenching equipment is mounted atop or across the pipe. A few systems can be employed in trench excavation for a pipeline on the seabed. For example, jetting is one of the systems, it is a post-laying trenching method which achieves a removal of soil below the pipeline by blowing seawater on both sides of the pipe with powerful pumps.

New words and expressions

1. submarine　adj. 水下的，海底的
2. uneven　adj. 不均匀的，不平整的
3. span　n. 跨度
4. bending stress　弯曲应力

5. vibration   n. 振动
6. subside   v. 沉降，沉淀
7. tie-in   接头
8. abrasion   n. 磨损
9. coating   n. 覆盖层，涂层
10. lateral   adj. 横向的，侧面的
11. sand ripple   沙纹
12. tensile   adj. 张力的，拉伸的
13. keel   n. （船的）龙骨
14. icicle   n. 冰柱
15. susceptible   adj. 易受影响的，敏感的
16. safeguard   v. 保护，保卫

Lesson 4.4   Marine disasters

Marine disasters refer to the natural disasters that occur in the ocean, which can result from the dramatic change of the normal marine environment. Marine disasters include the disasters from coastline to ocean. Like any other disasters, marine disasters are very dangerous to humans and have huge negative impacts on society, economy, properties, and life. Two typical types of marine disasters are described below:

Storm surge

Storm surge refers to the abnormal increase in seawater during a storm, and its size is measured by the height of water above the astronomical tidal level. Storm surges are mainly aroused by violent atmospheric disturbance due to strong wind or sudden change in air pressure. The amplitude of a storm surge is dependent on the relationship between the coastline location and the storm path, the intensity and size of the storm, the speed of storm movement, and the seabed topography.

The pressure effect of a tropical cyclone results in the seawater level rise in the region with low atmospheric pressure and fall in the region with high pressure. The water level rise counteracts the low pressure and keeps a constant total pressure at a certain plane below the water surface. The sea level is estimated to increase by 10mm for a mbar (hPa) decline in atmospheric pressure.

Besides the aforementioned processes, the nearshore surge as well as wave height are also influenced by the water flow over the bottom topography, i.e., the seabed morphology and elevation.

Factors affecting storm surge

Storm surge is generated by wind action rotating along the center of the storm to push the water towards the shore. Compared with the wind blowing to the shore, the low pres-

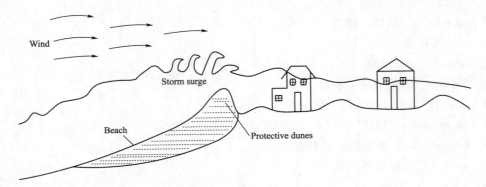

Figure 1 The process of a storm surge

sure associated with the strong storm has minimal contribution to the surge.

The potential maximum storm surge at a specific position relies on a few elements. Storm surge is a very complicated physical process in that it is closely related to the storm intensity, the moving speed, the size (radius of maximum wind speed), the angle approaching the shore, the central pressure (minimal influence comparing to the wind), and the features of coastal morphology.

Other elements affecting the storm surge include the width and slope of the continental shelf. A mild slope is prone to generate a larger storm surge in comparison to a steep shelf. For instance, a Category 4 storm that hits the Louisiana coast with a very wide and shallow continental shelf could generate a storm surge of 20ft high. However, if it is landed on the Miami coast of Florida with the shelf dropping off very fast, there may be only an 8ft or 9ft surge.

The destructive force of a storm surge is enhanced due to waves, and the continuous impact of waves on coastal buildings may increase the damage to coastal buildings. The density of water is $1t/m^3$ ($1700lb/yd^3$). The frequent action of waves on coastal structures is able to lead to the destruction of structures if the force is not specifically considered in the design. Surges and waves work together and enhance their onshore impact, in that the rise of water level leads to an inland extension of wave propagation.

In addition, the combined effects of current induced by winds and waves can lead to severe erosion of beaches and coastal highways. Buildings affected by strong storm are likely to be damaged by foundation erosion.

Tsunami

Tsunamis are a series of long waves caused by earthquakes, seabed volcanic eruptions, undersea landslides, the ejection of gases from volcanic mud, the collapse of icebergs, nuclear explosions, and even objects falling from the space, due to sudden and impulsive changes in water body or seabed.

The speed of the tsunami depends on the water depth. In the deep sea, tsunamis can move at a speed of 500 to 1000km/h. In coastal shallow waters, the speed is reduced to just a few dozen km/h.

The height of the tsunami also relies on the water depth. In a deep sea, the tsunami amplitude is generally only one meter high, but near the coastline, it can be as high as tens of meters.

Formation

Tsunami, often mistakenly considered as tidal waves, can be triggered in various ways. Seabed uplift (earthquake) in specific areas is the most common cause of tsunamis. Such tsunamis can travel thousands of miles before they disappear and often cause massive damage. Underwater landslide happens as a substantial mass of sediments moves along the seabed, which is another common source of tsunami generation. Landslides often produce tsunamis that are relatively localized, and they usually cause less damage than that by earthquakes. Finally, volcanoes and local landslides in coastal areas are also able to trigger tsunamis. However, these tsunamis occur in very confined areas and normally affect only a core region.

The "lifetime" of a tsunami consists of three phases: production, propagation and runup. Among them, we are most concerned with the first two. In a tsunami warning system, we try to do our best to monitor tsunami generation and understand the direction of tsunami propagation after they are generated.

The magnitude of a tsunami depends largely on two elements: the seabed topography and the tide. The undersea topography changes a tsunami wave height by modifying the ratio between the tsunamic wave length and its wave height. Generally, the ratio of wave length to wave height drops when the wave enters the shallow water, resulting in an increase in the tsunami size. The tide can also affect the damage to an area due to the tsunami, a high tide can cause much more destruction than that of a low tide.

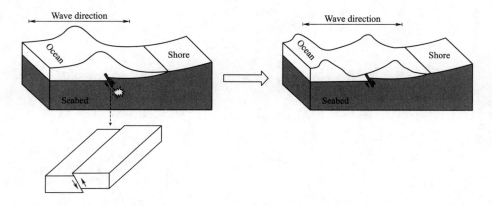

Figure 2　The generation and propagation of tsunami waves

Chapter 4 Ocean Engineering

New words and expressions
1. marine disasters   海洋灾害
2. dramatic   adj. 剧烈的
3. storm surge   风暴潮
4. counteract   v. 抵消
5. tsunami   n. 海啸
6. volcanic eruption   火山爆发
7. iceberg   n. 冰山
8. monitor   v. 监测，监控
9. destruction   n. 破坏，毁灭

课 文 翻 译

第 1 章 港 口 工 程

1.1 港口规划与布置

总体规划

港口可分为人工港口和天然港口。人工港口是指那些沿着海岸线通过填土或挖掘的方式建造的港口（图 1）。如图 1（a）所示，港口陆地部分是通过填土建造的；如图 1（b）所示，港池是通过开挖靠近海岸线的陆地人工建造的。挖入式港池的形状取决于港口的大小和开挖方式。挖入式港池通过进港航道与外海相连。为了防止波浪和海流带来的不利影响，通常会在港池的入口修建防波堤。

选择天然岸线建造港口时，需要考虑土地的可用性、填土材料、土壤质量、水深和环境条件等重要因素。

在进行海港工程的总体规划时，应提出初步设计标准，包括如集装箱、转运流程、进出口、船型设计和操作设备等的设计。现代货港更像是联合运输系统内的货物装卸枢纽，而非海上运输终点站。因此，港口的内陆连接是港口顺利运行和发展的一个基本要素。货物的转运可以通过与公路或铁路、人工或天然水路、航空线路或石油产品管线的连接来实现。

总平面布置

港口工程的布置应确保船舶容易靠泊、货物装卸安全、旅客下船安全。有必要通过设计合适的航道、港口入口和港池区域以及避免在港口区域内和周围产生有害的腐蚀或沉积物，来确保船只易于进入港口。

当船舶进入港池时，需要降低航速，以便进行锚泊动作。图 4 所示的是带有回旋水域的人工港口的平面布置。回旋水域可以位于港口之外，港口入口和主港口之间，或者位于靠近入口的主港池内。回旋水域又被称为调头水域，应根据港口设计船舶的尺寸进行计算。

防护工程总平面布置

港口防护工程的作用是确保港池和码头周围区域尽可能保持平静状态。港口防护工程主要包括以下建筑物：

1. 防波堤，可以与海岸线相连或者分离。与海岸相连的防波堤可以分为迎风式（或主防波堤）和背风式（或次防波堤）。前者保护港口免受主要来波方向的海浪的冲击，后者保护港口免受次要来波方向的海浪的冲击。一般来说背风式防波堤可以部分由迎风式防波堤来保护。

2. 突堤，通常成对设计。它们被布置在海岸线内侧或在河流中用于整治港口的入口，成对的突堤还可以增加流速来防止泥沙淤积。

港口工程通常位于破波带内，这里泥沙输运很活跃。通常在港口保护工程中，泥沙会淤积在迎风防波堤上游，侵蚀现象会发生在背风防波堤下游。

内港工程总平面布置

在工程区内，海床的岩土力学特性对工程的总体布局具有重要的影响。如果有岩石海床，通常建议将码头前沿布置在接近其设计深度的位置，以避免挖掘岩石所产生的昂贵费用。如果海床为软土地基，那么在码头位置的设计中，还要对填海和疏浚进行详细的技术和经济分析。

一般来说，总平面布置必须确保码头的形状能更好地利用港池，为船舶提供更方便的航行条件，并确保码头设备和机械的功能不受影响。此外，为了尽量减少港池的污染，应避免将码头建在水流较慢的区域。

码头长度是由具体的靠泊方法和泊位数量决定的。长度为 L 的船只停靠一般要求泊位长度为 $b=L+(30\sim40)$m 或 $b=1.2L$。码头前沿的最小水深 h 由设计船舶的最大吃水深度 d_{max} 由确定。还应该增加1m左右的安全系数，以考虑由于波浪引起的大幅度的运动，因此 $h \approx d_{max}+1$m。除系泊码头外，其他内部设施（如干船坞、滑道和维修码头）应独立于实体或透空式顺岸码头，并尽可能布置在受保护的港口区域。

1.2 码头功能分类

码头在功能上因其工作模式和运输商品的不同而不同。它们可分为集装箱码头、散货码头和件杂货码头。

集装箱码头

集装箱码头对劳动力的要求很低，但具有多种联运功能。然而，它们需要大量的储存空间，集装箱可以在这里用联运设备（起重机、跨运车和挂架）进行装卸。集装箱码头由于其联运功能，因而需要专门的起重机，如龙门式起重机和集装箱装卸桥。

集装箱是一种具有不同大小的可重复使用的箱子，比如，20英尺标准箱、40英尺标准箱和40英尺高箱。最常用的类型是20英尺长的集装箱，它常常被称为集装箱标准箱。它也是集装箱运输中的一种计量单位，也就是标准箱（TEU）。通常，一艘集装箱船，它的装载能力可以超过1000标准箱。集装箱不仅可以用于适箱货物的运输，也可以用于液体和冷藏货物的运输。

集装箱码头是在不同运输工具之间转运集装箱以便进一步运输的地点。转运可以在集装箱船和陆地运输工具之间进行，例如火车或卡车。在这种情况下，它被称作是海运集装箱码头。转运也可以在陆地运输工具之间进行，通常是在火车和卡车之间，在这种情况下，它被称作内陆集装箱码头。

集装箱码头往往是较大港口的一部分。它们通常为满箱和空箱提供储存设施。满载的集装箱储存时间较短，等待继续运输，而空载的集装箱可以储存较长时间，等待下次装载。集装箱通常被堆放起来储存，由此产生的储存区被称为集装箱堆垛。

散货码头

散货是指大量的无包装且尺寸一致的货物。液体散货包括可通过泵、软管及管道运送的原油和精炼产品。液体散货的运输使用数量相对有限的装卸设备,但需要大量的储存设施。干散货包括矿石、煤和谷物等多种产品。干散货需要更多种装卸设备,比如使用专门的抓斗、起重机和皮带运输机系统。对于特定的散货,可能需要对其特性进行一些改变,以确保运输过程的连续性,例如改变其装载单位或物理状态(从固体到液体或气体,或任何组合)。

散货码头是一种工业设施,用于货物的转运以及货物被转移到另一个设施进行加工或交付给最终用户之前的大量储存。码头的设计需要考虑三个主要方面:要储存的产品、进出运输方式和吞吐量。

散货码头通常由大型管道、装卸设备、储存设施和加工设施组成。让我们来看几个关于散货码头为多种行业提供关键基础设施的示例。

原油

油码头对能源公司的发展非常重要。大型油船向码头运送原油。原油被储存在地上或地下的油罐里。然后,原油被输送到炼油厂。码头和炼油厂通常建得很近,以便快速运输石油进行加工。

谷物

散货码头对谷物行业的发展也很重要。谷物通常通过船舶、铁路或卡车运送到这些码头,然后再转运出去。这些码头具有大的存储容量,能够联合火车和内河驳船更有效地运输谷物。

水泥

散货码头也为水泥提供了重要的配送渠道。码头使得通过驳船、火车或卡车运送水泥成为可能。这些不同的运输方式为满足不同客户的需求提供了灵活性。例如,驳船是快速运输大量水泥的一种经济高效的方式。火车能够长途运输水泥,而不用担心交通问题。卡车运输为不断增长的市场提供了快速而灵活的选择。

件杂货码头

件杂货是指具有多种形状、尺寸和重量的货物。因为货物是不均匀和不规则的,通过机械化运输是很难的。件杂货的搬运通常需要人工。因为货物的不同形式,件杂货码头需要同时配有堆场和仓库。

随着世界海运行业的发展,越来越多的货物将通过集装箱进行运输,这大大压缩了件杂货运输市场。但是由于件杂货自身的特性,在可预见的未来,集装箱运输还不能完全取代件杂货运输。因此,为了更好地建设低成本、高效率、高收益的新型件杂货码头,在码头的初步规划和设计过程中要发挥创造力,制定出最有效的设计方案。

1.3 码头结构分类

透空式顺岸码头、实体式顺岸码头、栈桥式突码头、实体式突码头

码头建筑物的主要功能是提供船舶安全靠泊的垂直前沿。根据码头建筑物的平面布置型式，它们可以被分为透空式顺岸码头、实体式顺岸码头、栈桥式突码头、实体式突码头。其中，透空式顺岸码头和实体式顺岸码头是平行于岸线的，栈桥式突码头和实体式突码头是垂直于岸线的。同时，透空式顺岸码头和栈桥式突码头一般建在桩上，而实体式顺岸码头和实体式突码头则是实体结构。

根据它们的建造形式，码头建筑物可以被分为多种类型。这里介绍了三种最常用的型式，即重力式码头、板桩码头和高桩码头。

重力式码头

重力式码头由基础、墙体（重力墙）、胸墙、护舷、滤层和墙后回填组成。重力式码头通常是用石块或混凝土在干燥的条件下建造的，前提是建筑地点可以排水且地基足够坚固。当地基相对较弱时，重力式码头可以在干燥的地方建在桩基上。根据结构设计的不同，重力式码头可细分为重力式方块码头、沉箱码头等。

重力式方块码头

重力式方块码头是最古老的重力式码头结构类型之一。它们由一个接一个的大块体组成，呈砖墙模式。这样的结构是用优质的天然石块或混凝土块在坚固的地面上建造的。它们的耐久性好，只需适度的维护。由于开采天然石块很困难，目前只考虑使用经济的混凝土块。墙体前的铺石的腐蚀是一个严重的问题。因此，许多老旧的方块结构不得不采用抗腐蚀的技术进行保护。

沉箱码头

对于沉箱码头，结构前部是由排成一行的预制的混凝土沉箱来建成的，与规划的新泊位的位置相对应。沉箱的形状和设计可能会有所不同，取决于不同的现场条件和可用的施工技术。矩形沉箱是最常见的类型。在沉箱码头中降低沉箱底脚外缘的应力比在方块式码头中更有效。增加沉箱的宽度，或将其分为两个或三个舱室且只在尾部的舱室压载，可以降低应力。沉箱的设计也必须使其能承受在生产、下水、拖曳、放置和压载过程中发生的荷载和应力。

板桩码头

板桩码头是港口工程中经常使用的码头结构类型。已经证明，对于墙体高度不超过 $18\sim20m$，且当地土质允许打桩的滨水建筑，板桩结构是一种可行且经济的方案。这些墙通常是柔性的，根据板桩的材料（例如，木材、钢或混凝土）、板桩支撑的方式和墙的施工顺序来分类。板桩码头的主要构件包括板桩、帽梁、拉杆、导梁和锚定结构等。

板桩码头是由板桩沉入地基来建造的。它的结构简单，因此减少了原材料的用量。同时，板桩码头的建造速度快且方便。然而，板桩结构的耐久性不如重力式结构。在板桩码头的施工期间，应避免承受大的波浪力。

高桩码头

在水深较小、地基承载能力有限的小型露天码头建筑中，混凝土桩或木桩通常用来做支撑。这种结构形式被称为高桩码头。如果平台需要承受大的荷载，这些桩必须紧密放置在一起。

如果基础的厚度超过 $3\sim4m$，以及如果地基处理需要大量潜水员和水下作业，则可使用横截面尺寸较大的桩。这些桩可以是混凝土桩，也可以是填充混凝土的钢管桩，或者钢筋混凝土桩。与混凝土桩相比，钢管桩可以被用于更坚硬的地层。例如，直径 70cm 的钢管桩可以打入由直径达 50cm 的石头组成的 20m 厚的碎石填土层。直径 50~80cm 的桩是最常用的桩型。

梁板式是高桩码头常用的一种结构形式。挪威建造的第一座钢筋混凝土码头，采用了高而窄的矩形梁来支承板。在使用 10~15 年后，人们发现这些结构梁底部钢筋出现了腐蚀的情况，以及随后覆盖钢筋的混凝土出现开裂和剥落。为了克服这些缺点，人们发明了无梁平台，它被证明非常耐用。然而，这种结构的建造成本更高，特别是因为需要模板支撑系统，所以现在又用回了梁板式结构。现代梁板式结构与老式的结构不同，它们具有低梁和宽阔的梯形截面。因此，避免了旧梁结构所经历的大部分问题。如今，梯形截面是露天码头结构中梁横截面的常用形状。

1.4 荷 载 分 类

在建造码头时，需要考虑的最重要的因素之一是结构所能承受的荷载。在建造之前，必须仔细分析特定设施的荷载要求，以确保可以安全地处理预期的交通流量。

固定荷载

码头设施设计应首要考虑其固定荷载。固定荷载是整个结构的重量，其中包括所有永久性的附属设施（例如灯杆、公用舱室、系泊构件、公用管线、保险库、工棚和平台）。为了规划设施的整个使用周期，对当前的和将来的附属设施都应当进行分析。进行这些评估时，应考虑实际可行的建筑材料的重量。

对于固定式的码头，通常是活荷载和横向荷载作为其控制变量。因此，高估固定荷载通常不会对结构的总成本产生负面影响。然而，高估浮体码头的固定荷载则会导致成本增加。

土压力也可能影响码头设施的设计，可能作用在固定结构上，从而导致结构横向移动。

垂向活荷载

垂向活荷载（或称移动/可移动荷载）是在码头设施建设中需要进行分析的关键因素。在结构上发生的特定荷载类型将在结构的设计中发挥重要作用。

铁路起重机的荷载在码头中很常见。许多起重机，例如集装箱起重机，有各种起重量、配置和尺寸可用。因此，没有来自起重机制造商的具体信息，就不应进行最终设计。尺寸的增加将导致起重机的自重更大，轮压也更高。作用在面板、轨道梁和桩帽上的最大轮压应增加 20%。

在开放式和浮体式的码头上，通常使用卡车式起重机。这些车辆在操作上有很大的限制。因此，应根据卡车式起重机的类型指定其实际荷载。

荷载可以通过几种方式来确定。对于集中荷载，例如来自具有轮胎的设备（卡车、卡车起重机、叉车等）的车轮荷载和悬臂浮动荷载，应将荷载施加到设计中考虑的导致最大作用力的位置和方向。通常，均匀和集中的活荷载应按逻辑方式施加，以避免在同一地区的来自轮胎式设备和轨道的活荷载同时发生。

在设计诸如面板和沟槽盖等短跨度结构时，重要的是要考虑集中荷载。其中包括卡车、叉车、移动式起重机和跨运车的荷载。对于梁、桩帽和支撑桩的设计，均布荷载是更重要的考虑因素。

水平活荷载

与垂向活荷载一样重要，设计者在建造码头设施之前必须考虑码头的水平活荷载。水平活荷载在确定一些构件类型方面起着重要作用，这些构件用来分散作用在结构上的力并保护结构在长时间内不受破坏。

码头的主要功能是船舶的停靠和系泊。靠泊荷载可能很大，尤其是在借助两艘或更多拖轮将船舶驶入时。船舶和码头设施之间的护舷系统可以减少靠泊能量和传递到结构上的力。传递的力的大小和位置将取决于许多因素，包括结构的类型、船舶的类型、靠泊速度和角度以及所使用的护舷系统。

一旦将船舶系泊，就会有作用力施加在码头设施上。这些力可能由风、海流和波浪导致。确定系泊荷载涉及许多因素：风、海流、波浪的方向和强度、系泊点的间距、泊位的暴露程度和船舶的方向、系泊缆的布局、系泊缆的弹性以及船舶的负载条件。如果码头位于有掩护的水域，则一般不考虑波浪力。但是，对于所有系泊的船舶，水流和风都会作用在其上产生作用力。计算机模拟程序通常有助于模拟潜在的系泊荷载。在设计码头结构时，施工过程中的系泊力也应被考虑。在多泊位码头中，结构本身应能够在最大风速下将船固定于泊位中。

由于潮汐运动或地下水积聚，水位差也可能对码头设施产生压力。例如，在板桩码头上，水压力会作用于板桩两侧。板桩上的实际受力是两侧的合力。

其他因素也可能影响码头结构上的荷载，例如浮冰。通常来说，与相同大小和厚度的海冰相比，淡水冰所产生的压力要小。

荷载组合

在分别考虑每个荷载因素之后，应分析各种荷载类型的组合，以构建能够承受面板上的交通、系泊的船只以及各种作用力的码头设施。这就需要考虑可能作用于每个结构和基础构件上的荷载组合。

确定码头设施的负荷要求是一个复杂的过程，它需要使用计算机模型和预测有关设施的预期用途。通过使用最佳估算并规划最大可能的荷载，设计者才能够确保码头结构拥有长期使用寿命。

1.5 货物装卸设备

货物性质和包装类型通常决定了货物装卸设备的选择。

在码头或港口，货物通过多种装卸设备在船舶两侧、中转仓库、仓库、驳船、铁路或公路车辆中运送。这些设备包括两轮手推车和四轮卡车。它们有些是手动的，有些是机械驱动的。也有由机械或电力驱动的拖拉机拖拽四轮拖车的。也有机械或电力驱动的皮带机，从码头前沿延伸到中转库、仓库、铁路或公路车辆。

在码头上有各种类型的岸壁式起重机、水平变幅起重机和移动式起重机等用于移动和提升货物。垂直方向的货物移动也是使用起重设备（吊上/吊下设备）完成的。

集装箱装卸设备

集装箱是运输非散货最常见的方式之一，因为集装箱可以在船舶、卡车和火车之间无缝移动，简化了整个物流链。事实上，在过去的几十年里，集装箱化的货物比例有所增长，达到了非散装货物的90%。集装箱化的货物需要非常专业的装卸设备和精心规划的码头配置，以达到以下目标：

- 确保以最经济的方式在正确的地点和时间交付正确数量的货品来提高效率。
- 减少物料在运输和储存过程中的损坏。
- 通过优化货物的存储配置，最大限度地利用空间。
- 减少货物装卸过程中的事故。
- 减少对环境有影响的排放。

集装箱起重机，也称岸壁式集装箱装卸桥，是在岸边装卸集装箱的。在码头移动和堆放集装箱的工具包括移动式集装箱吊运车、堆垛起重机、龙门式集装箱起重机和集装箱卡车。集装箱在码头内最多堆叠五层高，即五个集装箱一垛。用于装卸集装箱货物的工具包括叉车和卡车等。叉车常采用机械或电力驱动，其前部安装了一个由两个叉齿组成的叉型平台，起重能力从1~45t不等，一些叉车上配备有卷轴和捆束的夹具。

集装箱码头有三个主要的操作子系统：水侧——从船舶到码头的水域；堆场——存放集装箱的地方；以及联运——将集装箱运往内陆。

合适装卸设备类型的选择取决于码头的大小、平面布置、空箱与满箱间的空间分布和作业上的灵活性。

散货装卸设备

对于干散货，装卸设施可以使用皮带机，通常在陆地侧的一端由一个料斗或抓斗来装卸，它们对矿石可能是有磁吸力的，固定在大容量的起重机或龙门吊上。这些龙门吊架在码头平行方向移动，而且也能向岸移动一定的距离，这就覆盖了很大的堆场区域。这类设备可用于处理煤和矿石。当装卸散装糖时，糖会被卸下装进料斗，通过重力向轨道式卡车或公路车辆输送。筒仓通常用来装卸谷物。通过气动抽吸装置可以把谷物从船舱里吸出来。

液体散货的运输，如原油和精炼产品，是由管道从储油液舱运输到岸上储油库的。泵送设备安装在储油液舱或岸上的精炼厂上，而不安装在码头上。鉴于该类货物的危险性，一般做法是在离主码头系统较近的靠海侧建造专用泊位。货物油通过货物管道系统从船的储油液舱输送到船的总管，总管通常位于左舷或右舷。再通过岸上的装卸臂，转移到岸边总管，然后分配到石油码头的储油仓中。装卸臂软管必须密封，以避免漏油。

件杂货装卸设备

对于件杂货（商品、日用品等），它由码头上的门座起重机、浮式起重机或货船自带的装卸设备（甲板起重机、吊杆式起重机等）装卸。

许多类型的工具或可拆卸部件可以添加给船上或岸上的起重设备。常见的可拆卸部件包括吊索和绳带。这类设备一般由绳索制成，适用于吊装坚固的包装货，如木箱或袋装货物，它们在吊装时很难发生凹陷或损坏。同样，帆布也适用于袋装货物。吊索适用于重而细的货物，比如木材或钢轨。罐钩或桶钩适用于提升桶或鼓状货物。货网适用于邮件袋和类似的货物，它们起重时不容易被压碎。重型起重梁被设计用于火车头、铁路客运车厢和其他重而长的物品。吊货盘和托货板（后者是用木材或钢制作）十分适用于方便堆放的中等尺寸货物，如纸板箱、袋子或小的木质板条箱。

第 2 章 航 道 工 程

2.1 河 道 类 型

河道有三种基本类型：顺直型、蜿蜒型和分汊型。用以上术语来描述某一河道并不意味着整条河流都是顺直的或其他样子的，而是指河道的某些部分可以这样描述。事实上，一条河流的某些部分可能是顺直的，某些可能是蜿蜒的和（或）分汊的。

顺直河型

在传统分类中，顺直型河流一般被认为是平原河道地貌的典型类型之一。顺直型河道不太稳定，一般沿断层和节理发育。在自我调节的冲积河流中，极少数顺直河流分布在较大的时空跨度内。因此，问题就产生了：一条河道是否可能是顺直的？控制河道形成和由顺直型向其他形态转化的主要因素是什么？各种假说和理论，如地貌阈值假说、能量耗散极值假说和稳定性理论都可以回答上述问题，但不能解释顺直河道的形成。从现有的冲积平原形态来看，顺直河道不像自然界其他典型河道那样稳定。从河流沉积历史来看，没有明显的证据支持稳定的顺直河型的演化。

尽管自然界很少有河道完全是直的，但对顺直河型的描述是显而易见的。蜿蜒河型是沿河流方向迂回曲折的河道。地球科学家使用蜿蜒度来区分顺直和蜿蜒河型。蜿蜒度是指两点之间的河道实际距离与两点之间的直线距离之比。一条河流的蜿蜒度超过1.5，则被认为是蜿蜒河型。

蜿蜒河型

河曲是指河流、溪流或其他水道中出现一系列曲线、弯曲、弯折或曲流环，它是河流或水道在滩地上或河谷中从一侧摆动或移动到另一侧的过程中所产生。河曲是水流侵蚀凹岸沉积物并在下游凸岸淤积所造成的，其结果是河道在滩地上沿中轴线向下游滩地移动，形成弯曲的河道。曲流带指的是蜿蜒河道在滩地或谷底时常改道摆动的区域，其宽度通常是河道宽度的15至18倍。蜿蜒河道可以随着时间向下游移动，有时在短时间内发生显著变化，导致出现道路和桥梁维护这样的市政工程问题。

蜿蜒河型是在不同的自然条件和过程的影响下形成的。河道的波形结构是持续变化的。河水在弯道中以涡流的形式流动。一旦河流开始以正弦路径流动，螺旋流就会向弯道的凸岸输送大量的沉积物，同时弯道凹岸得不到保护，容易受到更严重的侵蚀，曲流的振幅和凹度会急剧增大，最终形成一个正反馈循环。

河道底部的横向水流将大量的沉积物带到弯道内部，之后向上流到河流凸岸附近的水面，然后流向凹岸，形成螺旋流。河道流速越大，曲率越大，则横向流越强，冲刷越剧烈。

在角动量守恒的前提下，弯道凸岸一侧的流速比凹岸一侧要快。因为水流的速度降低，水的离心力降低。在这种情况下，较高水柱体的压力占主导地位，形成不平衡的梯度，驱动河道底部的水从凹岸侧向凸岸侧回流。这种二次流将沉积物从弯道的凹岸侧带到凸岸侧，逐渐使河流变弯曲。

分汊河型

分汊河道由被一个或多个江心洲分隔开的相互交织的河道组成。分汊河型常出现在有高含沙量和粗颗粒河床质的河段，以及河流比降大于典型顺直和蜿蜒河型的情况，还跟河流水量变化迅速且频繁，以及河岸易受侵蚀有关。分汊河道分布在世界各地的多种环境中，包括以砾石为主的山区河流和流经冲积扇、沉积平原或河口三角洲的砂质河床的河流。

分汊型河流通常在有易被侵蚀的砂质河岸且很少植被保护的河段上可以找到。河床质较粗且不均匀。分汊河段的底坡比邻近非分汊河段的底坡要大。从水力学的角度看，分汊河段不如非分汊河段有效。分汊河段中，各支流的总宽度可以是不分汊河道宽度的 1.5～2 倍，水深则相应地较小。分汊是当河道变陡时的一种消能方式，原来由于流速增加将导致侵蚀的情况因此得以避免。

2.2 河床演变

河床演变是指自然情况下及修建整治建筑物后河床发生的冲淤变化过程，包括泥沙运动以及河床侵蚀或沉积。

平原冲积河流的河床演变主要体现在河槽中成型堆积体的发展和变化上。成型堆积体的变化不仅表现在淤积长大和冲刷变小，而且还会发生平面位移，使河槽的平面形态也发生变化，河岸在有些地方会蚀退，而在另一些地方则会淤长。

平原河流的河床演变与河型关系甚大。不同的河型具有不同的演变规律及形成条件。顺直型和蜿蜒型河段河道主要是边滩这种成型淤积体，其中顺直型河段的边滩呈犬牙交错状在河道两侧分布，而蜿蜒型河段的边滩则依附于凸岸。分汊型河段和游荡型河段的河道中具有江心洲或滩，其中分汊型河段江心洲稳定，而游荡型河段的江心滩不稳定，时常迁徙。冲积流的河槽形式在很大程度上取决于河床和河岸的相对易侵蚀性。

不论是试验水槽中或在野外，如果河岸质易被侵蚀，则水流流过初始的顺直河道时会形成曲流。蜿蜒型河道的长度可达非蜿蜒型河道的 1.5～2 倍。它的底坡相应地减小，但因河道变长和弯段损失，水头损失增大了。如果没有这些损失，流速就会较高，相应地产生下切河槽的趋势。但许多蜿蜒型河流，因下泄具有固定高程的水体，并不能下切河槽。若下切不能出现，就需要另外的一些机理来消除所具有的能量。

于是，分汊和蜿蜒都可以看作消能的方式。分汊多发生在河床质粗大且不均匀和河岸易被侵蚀之处。蜿蜒则可能发生在底坡较平坦、沉积物较细并且河岸质较有黏性之处。

影响河床演变的主要因素

影响河床演变的主要因素可概括为进口条件、出口条件及河床边界条件三个方面。

进口条件包括河段上游的来水量及其变化过程，河段上游的来沙量、来沙组成及其变化过程，以及与上游河段的衔接方式。

出口条件主要是出口处的侵蚀基准面条件。它可以是能控制出口水面高程的各种水面，如河面、湖面、海面等，也可以是能限制河流下切的抗冲岩层。侵蚀基准面情况不同，河流纵剖面的形态、位置及其变化过程会出现明显的差异。

河床边界条件泛指河流所在地区的地理、地质条件，包括河谷比降、河谷宽度、河岸及河谷的物质组成，以及河道几何形态等。即使进口的来水来沙条件和出口的侵蚀基准面条件完全相同，不同的河床边界条件仍会带来不同的河床演变特点。

河床演变的根本原因

对于任意一条河流的某一河段，当进出这一特定河段的沙量不相等时，河床就会产生冲淤变形。如果进入这一区域的沙量大于该区域水流所能携带的沙量，河床将淤积抬高；相反，如果进入这一区域的沙量小于该区域水流所能携带的沙量时，河床将冲刷降低。

因此河床演变的具体原因尽管千差万别，但从根本上可归结为河道水流的输沙不平衡的结果，即河床演变是输沙不平衡的直接后果。

当外部条件，即进口水沙条件、出口侵蚀基准面条件和河床周界条件保持恒定不变，且整个河段处于输沙平衡状态时，河段的各个局部仍可能处于输沙不平衡状态。例如，沙波和成型堆积体的存在使近底水流交替出现加速和减速，泥沙在水流加速区发生冲刷，而在水流减速区发生淤积，其结果使整体上仍处于输沙平衡状态的河床在特定地点已处于输沙不平衡状态。

使河流经常处于输沙不平衡状态的另一个重要原因是河流的进出口条件经常处于变化之中。由于流域降水的时空分布不均匀，进口水沙条件几乎总在变化。至于出口条件，如果着眼点是前文提到的侵蚀基准面，其变化是很缓慢的；如果聚焦于水流条件的变化，如干支流的相互顶托、潮波对洪水的影响等，则可能产生很大的变化。河床边界条件通常是比较稳定的，但当边界发生变形之后，例如在航道整治工程建设后，也可能触发新的输沙不平衡。

2.3 疏 浚 工 程

疏浚是利用挖泥船设备，将处于水下的沉积物挖出和转移到其他水域或者陆地的一种工程形式。其目的是维持航道和港口的通航，并通过收集底床沉积物并将其运输到其他位置来协助海岸防护、填海造陆和海岸重构。疏浚还可以回收利用具有商业价值的材料，包括某些矿物或可作为建筑材料使用的沉积物，如沙子和碎石。

疏浚工程通常分为三类：基建性疏浚、维修性疏浚和临时性疏浚。在新地点进行的疏浚，以及在以前从未疏浚过的材料中进行的疏浚称为基建性疏浚，例如填海造陆建设机场和人工岛、开发新港口、加深和拓宽航道等。维修性疏浚是指定期进行的疏浚工程，以维持或改善现有航道。临时性疏浚是为了解决工程量小的疏浚任务，通常是在没有常驻挖泥船的河段上，临时利用邻近地区的疏浚设备来进行工作。

疏浚通常包含四个步骤：松动底床质，将底床质转移至水面附近，运输以及处置。挖泥船按其工作原理可分为水力式和机械式两种。

水力式挖泥船

水力式挖泥船的主要特点是其疏浚的沉积物呈悬浮形式,并使用泵系统抽吸和管道排泥。砾石和其他粗颗粒沉积物也可以用更大功率的水力式挖泥船清除。一些常见的水力式挖泥船如下:

直吸式挖泥船

作业时,直吸式挖泥船可以用一根或多根定位桩使其固定在某个位置。直吸式挖泥船通过浮管将沉积物输送到需要吹填的岸边。长距离的输送则需要在管线中设置额外的增压泵。沉积物也可以直接装载到停泊在旁的驳船上。

绞吸式挖泥船

绞吸式挖泥船的抽吸管在吸入口配有绞刀装置。绞刀使底床沉积物松动并将其输送到吸入口。疏浚物通常由耐磨离心泵吸出,通过管道或驳船排放。绞吸式挖泥船常用于有坚硬底床物质的区域,如砾石沉积物或地表基岩。如果功率足够大,绞吸式挖泥船可以代替水下爆破。

耙吸式挖泥船

耙吸式挖泥船是一种自行推进的船,在挖泥过程中按照预先设定的路径填满其船舱或料斗。料斗可以通过底部或阀门倾倒,或通过泥泵将沉积物运载到岸上。这种挖泥船主要用于开放水域,如河流、运河、河口和开阔海域。

耙吸式挖泥船的外观呈普通船只形状,具有良好的适航性,无须任何形式的系泊或定位桩即可作业。用来衡量耙吸式挖泥船的标准是料斗容量,能在几百立方米到超过 2 万 m^3 范围内变化。近年来建造的越来越大的船只使疏浚物的运输越来越经济,尤其是用于填海造陆工程时。

机械式挖泥船

机械式挖泥船有各种各样的形式,但每一种都有相同的"抓取"工作原理。这些设备都配有抓斗或铲斗来抓取松散的河床沉积物,然后将物料填满在铲斗中,再提起铲斗将其运送到倾倒场。一些常见的机械式挖泥船简介如下:

链斗式挖泥船

链斗式挖泥船是对传统斗式挖泥船的改进。它们可以处理各种各样的床面物质,包括较软的岩石和珊瑚。但由于其效率低,噪音大,而且需要锚线,因此链斗式挖泥船的使用近期已经大为减少了。

铲斗式挖泥船

铲斗式挖泥船和挖掘机一样有一个反向铲斗。在浮码头或驳船上配备一个在地面使用的反铲挖掘机即为一个简易但可用的铲斗式挖泥船。小型铲斗式挖泥船可以安装在挖槽岸边的轨道上作业。疏浚物通常装在驳船里。铲斗式挖泥船主要用于港区及其他浅水区。

抓斗式挖泥船

抓斗式挖泥船用船载起重机或浮吊上的蛤壳式铲斗抓取底床沉积物。该设备常用于海湾中挖泥。这些挖泥船大多是带有定位桩和升降钢桩的起重机驳船。

2.4 整治建筑物

整治建筑物是一种调节河道的水利工程，包括丁坝、顺坝、锁坝、生态护岸等。整治建筑物包括重型或大型建筑物，它们通常是航道整治总体规划中的一部分，并为长期使用而设计；而轻型整治建筑物，主要在中小型河流上阶段性使用。

丁坝

丁坝是从岸边修筑伸向水中的建筑物，以保护堤岸不受侵蚀。这类建筑物被广泛用于航道整治工程项目，并具有以下一项或多项功能：

（1）丁坝通过导流、挑流和维持河道内的水流，使河流沿着预想的路线前进。

（2）丁坝能形成缓流区，以促进丁坝附近区域的淤积。

（3）丁坝能防止主流冲刷河岸，保护堤防。

根据不同目的，丁坝可以单独或联合使用。丁坝可以垂直于主流，也可以以一定的角度指向上游或下游。它们也可以结合其他整治措施使用。如果要保护的河段较长，或单一丁坝不足以有效地改变水流，并且对上下游沉积物的影响也不够明显，则可采用丁坝群。在丁坝群中，位于最上游的丁坝在其向河和向陆的两端都更容易受到水流的冲击。因此，为了确保其结构稳定应给予特殊处理。

通过作用于周围的水流，丁坝往往会增加其附近的局部流速和湍流强度。丁坝本身的结构可能易受到侵蚀。平行于河岸流动的水流被拦截后，沿着丁坝的上游面加速流向坝头。坝头附近水流流速和曲率增大会导致附近河床的严重冲刷。整治建筑物必须有足够深的基础或者足够好的防护，否则丁坝的坝头和坝根部分可能会受到局部冲刷的破坏。

丁坝与主流的夹角可能会影响整治结果。与河流垂直的丁坝通常是最短的，因此也是最经济的。上挑丁坝能更好地保护丁坝的向河端免受冲刷。下挑丁坝更适合于保护凹岸，尤其是如果丁坝的长度和间距能够使主流远离凹岸，从而给整个河段提供保护。

顺坝

顺坝是一种纵向布置的整治建筑物。坝身一般较长，与水流方向大致平行或有很小交角，通常沿整治线布置。它具有束窄河槽、引导水流、调整岸线的作用，因此又称作导流坝。顺坝常常布置在分汊河段的分流汇流区以及河口治理段。

锁坝

锁坝是一种拦截河流汊道的水工建筑物。为了达到增加航道水深，满足河流通航的要求，应该对部分狭窄的、不具备通航能力的或者通航危险的河道进行封堵。坝体两端嵌入河岸或江心洲，坝顶中部呈水平，两侧向河岸升高。

生态护岸

混凝土和岩石材料在传统的河堤护岸工程中广泛应用，以避免洪水和侵蚀带来的恶果。坚固的护岸以及稳定的结构对于确保人员及其财产的安全至关重要。然而硬质护岸会对生态系统产生负面影响，包括水生和两栖动物栖息地、河流水质和美学价值等方面。为了构建安全的护岸同时减少对河流生态系统的负面影响，必须考虑护岸工程的生态影响。

生态护岸必须将土木工程技术和河流的地形、水文、生态和其他条件结合起来考虑，以保证岸坡良好的稳定性和生态修复效果。生态护岸常使用多孔结构和植物，以促进地下水和河水的循环，促进河岸生态恢复，这也是"海绵城市"概念中不可缺少的一部分。生态护岸也可以使用石灰石材料建造。多孔结构的生态护岸还包括石质边坡防护、石笼防护、弧形石河床防护等，可为微生物和植物繁殖提供栖息地，对处理水污染具有重要作用。植物和微生物的生长及其多样性对河流水质也有积极影响。

2.5 船　　闸

船闸是一种有助于通航的设施。由于受天然河流中流量调节、渠化工程的影响以及运河上河床地形、水面坡度的限制，时常有必要创造沿流方向阶梯状的水位落差。所以必须使用专门的通航建筑物协助船舶平稳地通过这样的落差。现代通航建筑物应用最多的是船闸。它是由上、下游引航道和上、下闸首以及闸室组成。闸室是一种停泊船舶或船队的厢形室，借助灌水或泄水来调整闸室中的水位，使船舶在不同水位之间作垂直的升降，从而通过急剧的航道水位落差。当船舶或船队由船闸的下游向上游行驶时，闸室内水位降至与下游水位齐平，然后下游闸首的闸门开启，船舶或船队进入闸室，然后闸门关闭，闸室注水，待水位升高到与上游水位齐平后，开上游闸首闸门，船即可出闸通过上游引航道驶向上游。当船舶或船队由上游向下游行驶时，船闸操控程序则与此相反。

船闸最重要的一个特点是有固定的闸室，闸室内的水位可以升降变化，然而箱形船闸或升船机的升降运动则是船闸本身（通常称为沉箱）。船闸由一个矩形闸室组成，该闸室具有固定的侧面、可移动的闸门和用于输水和排水的设施。闸室的输水和排水可通过手动或机械操作的泄水闸来实现。闸室的尺寸取决于使用航道的船舶的最大可能尺寸。在航运繁忙的地方，可能需要两个或多个闸室。

船闸的主要构成

船闸由引航道、闸首、闸室、工作闸门和阀门、输水系统等部分以及相应的设备组成。

引航道

引航道是连接船闸和主航道的一段过渡性区域，分上游引航道和下游引航道，其平面形状和宽度、水深要能使船舶或船队安全迅速地进出闸室。引航道进出口处水流流向与流速要能满足船舶安全进入和驶出的要求，并防止泥沙由于回流的作用而淤积在引航道上。引航道内一般设有导航和靠船建筑物。导航建筑物为不透水的导航墙，紧靠闸首布置，用以保证船舶安全进出闸室。靠船建筑物供等待过闸的船舶停靠用。

闸首

上下闸首是将闸室同上下游航道隔开的挡水建筑物。闸首配备有工作闸门、检修闸门、输水系统、闸门和阀门的启闭装置等。闸首通常是整体式钢筋混凝土结构，它的边墩和底板刚性连接在一起。

闸室

闸室是由船闸的上、下闸首和两侧的闸墙围成的空间。闸墙上配备有系船柱、浮式系船环等，供船舶在闸室内停泊时系缆用。过闸船舶在闸室中随着闸室内水面变化而升降。

2.5 船　　闸

闸室结构一般采用砌石或钢筋混凝土材料，有两种设计方式：闸墙和闸底可以刚性连接在一起成为整体式结构，也可以不连接在一起成为分离式结构。

工作闸门和阀门

工作闸门是安装在上、下闸首的活动挡水设备。闸门在闸首两侧水位相同时开启和关闭。因为运转频繁，所以要求闸门操作灵活且迅速。人字形闸门应用最广，常用的还有平板升降闸门、横拉闸门、扇形闸门（又称三角闸门）等。输水阀门是设在输水廊道上，用来控制灌泄水的流量。它是在水压力作用下开启的，要求结构简单、启闭力小、操作方便。三类最常用的阀门有：平板提升阀门、蝴蝶阀门和反向弧形阀门。

输水系统

输水系统是给闸室灌水和泄水的设施。闸室的灌泄水时间应尽量短，并满足船舶停泊平稳的要求，一般为 10~15min。输水系统的基本形式有两种：①集中输水系统（又称头部输水系统）。闸室灌水、泄水分别通过设在上、下闸首内的输水廊道在闸首处集中进行；②分散输水系统。闸室灌水、泄水由输水廊道通过分布于闸室底板或闸墙内的出水口进行。如果水头（船闸上下游的最大水位落差）在 15m 以内，一般采用集中输水系统。水头较大时应当采用分散输水系统。

第3章 海岸工程

3.1 海 浪

波浪的产生

海浪主要是由风对水的作用产生的。波浪最初是由一个复杂的剪切和共振作用过程形成的，在这个过程中，产生不同波高、波长和不同周期的波并向不同方向传播。一旦形成，波浪可以传播很远的距离，在传播过程中，波高会减小，但波长和周期可以保持不变。

在风暴产生区，高频波（即短周期波）的能量被耗散或转移到较低频率。不同频率的波以不同的速度传播，因此在风暴产生区之外，海况会随着各种频率波成分的分离而发生变化。低频波的传播速度比高频波快得多，导致产生了涌浪而不是风浪。这个过程被称为色散效应。因此，风浪的特点是陡峭、不规则和短峰的，涵盖了一系列的频率和方向。相反，涌浪的特征是较平坦、相当规则和长峰的，包含频率和方向的范围较窄。

波浪的传播与衰减

波浪浅化

在流体力学中，波浪浅化是指进入浅水的波浪的波高发生变化的一种效应。这是由于波群速度，也就是波能传输速度随着水深而变化。在稳定的状态下，波浪必须通过增加能量密度来补偿能量传输速度的降低，以保持恒定的能量流，浅化波的波长也会减小但频率保持不变。

海床摩擦

过渡水深和浅水中的波浪会由于海床摩擦导致的波浪能量耗散而衰减。这种能量损失可以用线性波理论来近似计算，类似于管道和明渠水流的摩擦关系。与恒定流中的速度分布不同，波浪作用下的摩擦效应会产生一个非常薄的振荡的波浪边界层。因此，它的速度梯度远高于等效恒定流中的速度梯度，相应这也意味着波浪摩擦系数将大几倍。

波浪折射

它是一个行进的波浪与海底地形的相互作用而改变传播方向的过程。波浪在到达海岸线之前，在远海中传播数千米。通常情况下，它们会垂直于海滩到达海岸线附近，产生与岸线平行的波浪。然而，在世界上一些地区，当涌浪波前线以不同的角度到达浅水区时，它们往往会从深水区转向浅水区。这种现象的发生实际上是因为水深变浅减小了波浪的传播速度，而在深水区运动的那部分涌浪仍然以相同的速度继续运动。

波浪反射

当波浪传播到固体障碍物时，如防波堤、海堤、悬崖或斜坡岸滩，就可能发生反射。

对于垂直和硬质的结构，反射波能所占的比例可以很大。对于可渗透的结构或平缓底坡，反射要少得多。在发生反射的海岸的靠海侧，波浪运动是由入射波和反射波的叠加形成的。一个众所周知的叠加结果就是出现驻波，它是正向入射的单色波列在直立海堤上完全反射后出现。通常海岸的反射率小于100%，因此会产生一个部分驻波而不是完美驻波。

波浪绕射

当波浪在传播过程中遇到障碍物时，它会改变方向，或者绕过它。在海洋中，当波浪遇到像突堤这样的构筑物，并且波浪围绕着它旋转时，就会发生这种情况（有时当波浪穿过防波堤上的一个小开口或在两个岛屿之间运动时，也会发生绕射）。对于波长较长（即周期较长）的波浪，波浪的"缠绕"或转向效应会更大。绕射可以发生在浅水或深水中，并与折射有所不同，因为它不是水深变化导致的。然而，折射和绕射都会涉及波浪传播方向的改变。

波浪破碎

接近海岸的波浪会随着水深的减小而变陡。当波陡达到临界值时，波浪会发生破碎，消散大量能量，同时产生平均水位升高（称为波增水）。波浪破碎带是一个从初始破碎位置延伸到波浪最大上爬点的区域。波浪破碎可分为三种主要类型：崩破波、卷破波和激破波。

崩破波

当波浪在靠近海滩的平缓斜坡底床上传播时，会发生崩破波。波浪破碎过程缓慢且距离长，白色浪花从波峰沿着波前倾泻而下，同时损失能量。

卷破波

当波浪到达中等坡度的底床时，会产生卷破波。它比崩破波更陡，波峰以清晰的卷曲形式下落并携带大量的能量。当这类波浪以一定的角度冲击海岸线并穿过海岸线时，由卷曲水体形成的管状浪被冲浪者所喜爱。

激破波

当长周期、小振幅的波浪传播到陡峭的岸滩时，会出现激破波。波峰不会溢出或变卷曲，它会积聚起来，然后迅速垮塌到岸滩上，与其他两种破碎波相比，它产生的泡沫或浪花更少。

3.2 防波堤、海堤和丁坝

防波堤

防波堤是一个在海岸附近建造的建筑物，作为海岸防护的一部分，旨在保护港池或锚地免受恶劣天气和近岸流的影响。防波堤结构的建造是为了吸收海岸附近波浪的能量，可以通过使用砌块（如沉箱）或使用护岸斜坡（如块石或混凝土防护单元）来实现。一般而言，护岸是背靠陆域的结构，而防波堤是背靠水域的结构，即结构的两侧都有海。

抛石防波堤

抛石防波堤是利用结构的空隙来耗散波能。它是由一堆按单位重量分类的石头组成：小的石头作为核心，而大的石头则作为保护层，以保护核心免受波浪冲击。结构外侧诸如岩石或混凝土块体的防护单元吸收了大部分波能，而砾石和沙子阻止其余的能量穿透防波堤的核心。根据所用材料的不同，护坡的坡度通常在1：1至1：2之间。在浅水中，建造护坡式防波堤的费用通常更便宜。随着水深的增加，由于材料需求增加成本显著增加。

沉箱防波堤

沉箱防波堤通常有垂直的侧壁，并在船舶和其他水上交通工具需要停泊的地方建造。沉箱尺寸及其质量是为了抵抗入射波的倾覆力而设计的。在浅水中建造沉箱防波堤相对昂贵，但在更深的地方建造沉箱防波堤可以节省大量的费用。为了耗散波能，有时在沉箱前增加一堆抛石，从而减轻垂直壁上的波反射和动水压强。这种设计给防波堤的向海侧壁面提供了额外的保护，但它也增强了波浪在结构上越浪。

开孔墙式沉箱防波堤

吸波式沉箱是一种类似但更先进的设计，在它向海侧的壁面上开有不同类型的孔。这种结构已经成功地用于海上石油开采以及使用低顶高程结构物的海岸工程中。开孔墙式沉箱防波堤的主要优点是节省了在较深水域的施工成本，且对周围水环境阻碍较小。

浮式防波堤

与传统的固定式防波堤相比，浮式防波堤是保护海岸免受波浪作用的另一种选择。这种方法对波浪不大的沿海地区有效。因此，它们越来越多地用于保护小港口或码头，少数时候也用于减少海岸侵蚀。浮式防波堤一般分为四大类：箱式、浮筒式、毯式和系缆浮子式。

海堤

事实上，海堤和护岸是同义词。海堤被用来分割陆域和水域。它的设计是为了防止极端海浪或风暴潮造成的海岸侵蚀和其他破坏（如洪水），它的建造规模通常非常巨大。

海堤沿海岸线或在悬崖或沙丘的根基部修建。海堤通常是一种混凝土斜坡结构。它向海侧的表面可以是光滑的、阶梯状的或弯曲的。它也可以被建造成为一个抛石结构或一个混凝土、钢质或木质结构。海堤设计的一个特点是该结构能够承受强烈的波浪作用和风暴潮。抛石海堤与抛石护岸非常相似，通常用于保护非柔性海堤的根基部。然而，护岸通常被用作海堤的补充，或在受波浪影响较小的海岸上作为一个单独的结构存在。有时候，一个暴露于波浪中经过加固后可以抵御波浪冲击的堤坝也可以被视为海堤。

海堤是一种被动防护结构，它保护海岸免受侵蚀和洪水的侵袭。常被用于受波浪影响的城市海岸带，那里陆域面积稀缺因此需要良好的防护，通常在这些海堤的顶部会建造人行步道。海堤也被用于某些人口较少的沿海地区，特别是一些迫切需要海岸和海上防护相结合的区域。

丁坝

丁坝是一种从海岸延伸至海中的海岸构筑物，通常垂直或稍微倾斜于海岸线建造。为了达到保护海岸的目的，它通过促进泥沙淤积，削弱斜向波和沿岸流对海岸的侵蚀。

丁坝设计（平面形式、长度、高度、垂直于海岸的剖面形态、倾斜角度等）对海岸形态的影响较大，还与海平面、海浪条件以及破碎带内的泥沙输运有关。仅用一个丁坝来保护海岸是非常低效的。因此，一个典型的海岸保护工程通常采用由几个到几十个独立结构组成的丁坝群。

丁坝高度影响其拦截的沿岸输沙量。同一丁坝可以在非淹没或淹没任一情况下发挥作用，取决于潮汐或风暴潮导致的水位变化。通常来说，丁坝被设计成高出平均海平面 0.5~1.0m。如果丁坝太高，增加的反射波会导致局部冲刷。从平面上来看，丁坝可以是直的、弯的、L 型、T 型或 Y 型。

3.3 海 岸 地 貌

海岸地貌有两种主要类型：一种是侵蚀型，另一种是堆积型。它们表现出明显不同的地形，尽管每种类型都可能包含另一种类型的一些特征。一般来说，侵蚀型海岸是指沉积物很少或没有沉积物，而堆积型海岸的特征是长期积累了丰富的沉积物。

侵蚀型海岸的地貌类型

岬角

岬角是一种海岸地貌，通常是指陆地地势高的点，通过一个陡峭的下降延伸到水体中，尺寸相当大的岬角通常被称为海角。岬角的特点是地势高、陡峭的悬崖、岩石海岸、波浪破碎和强烈的侵蚀。

海蚀洞

海蚀洞是主要由于海洋波浪作用塑造的一种地貌类型，最相关的成因是侵蚀。在世界各地都发现了海蚀洞，它们沿着当前的海岸线形成，并在先前的海岸线上留下残存的海蚀洞。世界上最大的一些由波浪侵蚀的洞穴是在挪威海岸发现的，但它们比目前的海平面高 100ft 或更多。

海蚀拱

另一种壮观的侵蚀地貌是海蚀拱，是由不同速率的侵蚀而形成，通常是由于基岩的不同抗侵性所致。这些海蚀拱可能是一个拱形或矩形，其开口延伸到水位以下。一个拱的高度可以高于海平面几十米。

海蚀柱

海蚀柱是由海岸线附近位于海水中的一种陡峭的，甚至是垂直的岩石柱组成的地质地貌。它由波浪侵蚀形成，有些高达几米，并在波浪作用且比较光滑的那面形成孤立的尖峰。由于侵蚀是一个持续的过程，这些特征不是永久的。侵蚀最终将导致岩柱倒塌，留下一个残端。

堆积型海岸的地貌类型

沙坝

沙坝是被淹没或部分暴露的沙脊或由海浪从海滩向外海搬运的粗糙沉积物。海浪在海滩上破碎时产生的湍流在沙质海床冲刷出一条沟槽，其中一些沙被向前输送到岸滩上，其余的沉积在沟槽的离岸侧。回流和裂流中的悬沙对沙坝的形成亦有贡献，一些从更深的水向海岸移动的沙也同样如此。由于沙坝上存在波浪破碎的冲击，它的顶部通常维持在静水以下。

连岛沙洲

当远离河口时，在波浪影响较弱的地方也会发生沉积。例如，一个靠近海岸的小岛可以保护陆地海岸线不受大的海浪的影响，从而减少了岛屿和陆地之间的波浪活动。这个区域被称为波浪掩护区。沿岸流不能通过这块平静的区域，因为沿岸流的驱动需要波浪的运动。因此，大部分由沿岸流携带到这块区域的泥沙都会沉积于此。泥沙的沉积物最终在陆地和岛屿之间形成了一座沙桥，这种沙桥被称为连岛沙洲。

沙嘴

沿岸流携带的泥沙也会造成沙质海滩的一种延伸，被称为沙嘴。沙嘴会部分延伸到一个凹岸或海湾的入口。一旦沿岸流遇到凹岸或海湾时，漂移路径的改变会将沿岸流带到更深的水中。然而，随着水深增加，海床附近的波浪运动就会减小，因此漂流也无法维持，然后携带的泥沙会沉积起来，形成一个沙嘴。一旦沙嘴形成，沙嘴向海一侧的波浪作用仍然很强烈，沿岸流继续将泥沙输送到沙嘴的末端，并在那里沉积。沙嘴保护了海湾免受海浪的侵袭，在它向陆地的一侧提供了一个有掩护的平静水域。发生在掩护区沙嘴末端的沉积可能会驱使沙嘴向陆地方向偏转，并形成一个钩状结构。更进一步，当沙嘴发展到海岸时，可形成一个环状结构。

堰洲岛

陆地海岸线也可以被堰洲岛保护而免受海浪的影响，这些堰洲岛是与陆地海岸平行的长而狭窄的岛屿。通常存在一个浅湾，也被称为潟湖，将堰洲岛与陆地分开。一个堰洲岛很可能是由一个已经形成的沙嘴，或是一个后来被海平面上升所淹没的沙丘导致。由于风和/或水流的影响，堰洲岛通常处于迁移状态。

3.4 海岸输沙和海滩养护

海岸输沙

不同的海岸地貌与复杂物理过程的相互作用造成了海岸泥沙输运，也被称为近岸输沙，通常分为沿岸输沙和垂直于海岸方向输沙。

沿岸输沙

沿岸输沙很大程度上取决于波浪斜向海岸线发生破碎时所产生的沿岸流。沿岸流导致沿岸漂移现象，这个地质过程包括沉积物（黏土、淤泥、卵石、沙子等）沿着与岸线平行的海岸输运，取决于斜向来风的风向。该种风沿着海岸作用于水体，产生一股与海岸平行移动的水流。结果沿岸流驱动了沉积物的运动，也被称为沿岸漂移，这个过程通常发生在破波带内。

在那些斜风天里，由于海滩上海水的上冲和下泄，海滩的沙是在移动的。破碎波以倾斜的角度向上输送海水（上冲）而重力则导致海水垂直于海岸线下落（下泻）。因此，海滩上的沙可以以曲折的方式向远离岸滩的方向移动，一天几十米，这种现象被称为"岸滩漂移"，然而一些人认为它只是"沿岸漂移"的一种形式，因为沙子总体上的输运与海岸平行。

沿岸漂移会决定许多泥沙颗粒的尺寸，因为它的作用方式略有不同，这取决于沉积物的性质，例如沙质岸滩与砾石岸滩的泥沙在沿岸漂移方面的差异。泥沙在很大程度上受到近岸波浪往复作用的影响，因为破波和沿岸流的底床剪切力可以导致泥沙运动。由于沙质岸滩要比砾石岸滩平缓得多，卷破波更有可能在后者发生，由于缺乏一个延伸的破波带，大多数沿岸输运发生在冲流带。

垂直于海岸方向输沙

沿岸输沙主要是由于波浪产生的沿岸流造成，而垂直于海岸输沙则是由于波浪和海底回流引起的海水运动所致。季节性海岸线的变化通常被认为是对冬季发生风暴的一种反应，由离岸方向的泥沙输运和在离岸沙坝附近泥沙沉积造成。对于离岸沙坝形成的确切原因学者们没有共识；然而人们普遍认为是由破碎波导致的海底回流造成。风暴潮引起的波浪波高越大，沙坝形成地点离海岸更远。随着水深增大，沙坝的整体尺度也同样增加，需要更大量的沙，部分沙是通过侵蚀海滩剖面陆上部分提供的。

海滩养护

海滩养护是一种主要用于应对海岸线侵蚀的适应性技术，这类技术亦可能对减少海岸洪水有利。这是一种通过对存在缺沙情况的海滩人工补充合适数量的沙子进行海岸保护的软工程方法。有时候，海滩养护也可以被称为海滩回填、海滩补沙、重新养护和海滩喂养。

在海滩养护工程过程中，有两种主要的影响海滩养护效果（即养护过程中有效增加的海滩宽度）的物理过程，每个过程运行在不同的时间尺度上。第一个物理过程，即人工海滩养护过程中被人为改变的海岸地形构造，将通过在回填的海滩上发生的离岸输沙重新达到一个平衡地形，这个过程发生在数月到数年的时期内。第二个物理过程，则是被海滩养护工程人工改变的海滩岸线对沿岸流产生的扰动，导致回填沙在一个长期缓慢的沿岸输沙过程中被输运到回填海滩附近的其他地方。对于持续时间足够长的海滩养护工程，这个过程发生在数年到数十年的时间尺度上。

海滩养护工程的设计原则是顺应自然海滩的动力过程，让回填沙可以响应波浪和水位变化而不断地发生输运。根据当地的具体情况，海岸工程师可能会选择将回填沙直接铺在海滩上，或者堆在水中，或者垒成沙丘，或者三者兼有。这些回填的沙会随着海岸动力过程开始被重新输移，重新分布在海岸附近，从而改变岸线的响应规律。最终达成提升海滩宽度、调节海滩坡度和提升海滩岸线防护功能的目的。

经过养护工程处理的海滩坡度变得平缓，波浪向岸传播过程中通过波浪在浅水上破碎消耗更多的能量，从而减少了抵达岸线的波浪能。海底回流、沿岸流的存在形成离岸方向和沿岸方向的输沙，将一部分新回填的沙体带到较深的水域，或形成沿岸输沙。这些受离岸和沿岸输沙影响的沙体经常会形成一个离岸的沙坝，它将导致波浪在离岸更远的地方破碎，进一步消耗了抵达岸线的波浪能。

为了确保养护过的海滩能够在时有发生的风暴导致的极端海况下持续提供足够的海滩宽度和岸线防护，海滩养护必须周期性地进行。不断地向海滩补充新的沙体以供长期输沙消耗，这个过程称为周期性海滩养护。

第4章 海洋工程

4.1 海上平台

海上平台，也被称为石油平台，是一种具备在海洋深处钻井和开采石油天然气能力的巨型结构物。这些平台拥有专门的设施来储存原油和天然气，直到它们被运往炼油厂。有的平台还具备相应设施为劳动力提供住宿。根据需要，海上平台可以是浮动式的或固定于海底。

海上平台的主要类型

固定式平台

固定式平台是建立在大型钢或混凝土支撑腿上的，这些支撑腿直接安装在海底。固定式平台拥有足够的空间放置钻机和生产装备，并且还能为船员提供住宿设施。这种类型的平台稳定性高，并且为长期服役而建造。通常情况下，它可以安装在水深达520m（1700ft）的海域中。当水深超出这个范围，过高的成本会导致平台失去实际应用价值。

重力基础式结构（GBS）

GBS可以是钢制或混凝土制的结构，通常直接安装在海床上。钢制GBS主要用于没有起重机驳船，或起重机驳船不适用于安装传统的固定式海上平台的地区，例如在里海。全世界现存的钢制GBS的例子可在土库曼斯坦近海水域（里海）和新西兰近海看到。钢制GBS通常不具备储存碳氢化合物的能力。GBS的安装是通过湿拖或（和）干拖的方式把它从码头拉出来，然后通过控制压载水舱的海水来进行自安装。为了在安装过程中定位GBS，GBS可以通过绞索千斤顶与运输驳船或其他类型驳船相连（只要驳船尺寸足以支撑GBS）。在给GBS加载时，应慢慢释放千斤顶，以确保GBS只会产生偏离设计位置稍微一点的摆动。

顺应塔式平台

顺应塔式平台是基于固定式平台的设计思想来建造的。然而，它使用由混凝土和钢制成的窄塔式结构。这类平台被设计成具有柔性，可以在风和波浪的作用力下摇摆或横向移动。顺应塔式平台可以在457~914m（1500~3000ft）的水深变化范围内作业。

单柱式平台

在这种单柱式平台设计中，平台被安装在一个巨大的中空圆柱形船体上，圆柱的底端可以伸到水下约213m（700ft）。尽管圆柱体结构远离海床，但由于其圆柱体本身的质量，使得其上的平台仍然能待在原地。这类单柱式平台可在水深高达3048m（10000ft）的地方作业。

半潜式平台

顾名思义，这是一类半淹没式的平台，并且可以根据需要从一个地方移动到另一个地方。由于半潜式平台的下船体被水淹没，因此其受波浪荷载的影响要小于普通船舶。然而，由于半潜式平台水线面较小，其对荷载变化非常敏感，因此必须通过调整载重以保持稳定性。与全淹没式结构不同，半潜式平台没有靠在海床上的支撑。它们依靠动力定位的工作原理，并使用大型的锚来保持位置。这种类型的平台可以在 60～3000m（200～10000ft）的水深范围内工作。海洋石油 981 是第六代深水半潜式平台，也是中国第一台采用第三代动力定位系统的钻井平台。它的设计排水量为 30670t，平均吃水为 19m，总长度是 114.07m，宽度是 78.68m。

自升式钻井平台

自升式移动钻井装置（或简称为自升式平台），顾名思义，是一种可以通过降低支腿将钻井平台顶起的装置，就像千斤顶一样。尽管有些设计可以使装置的工作水深达到 170m（560ft），但这类装置一般还是用于 120m（390ft）以上水深的海域。自升式钻井平台能够从一个地方移动到另一个地方，然后可以通过每条支腿上的齿轮齿条系统驱动支腿伸向海底来固定自己。

海星平台

这类平台是半潜式平台设计的一种较大版本。然而它们是通过柔性的钢支腿连接到海床上，而不是锚。这类平台通常在 152～1067m（500～3500ft）的水深范围内工作。

张力腿式平台

张力腿式平台属于浮动式平台的类型，除了其特有的张力支腿可以从海底延伸到平台以外，它通常可以看作海星平台的巨型版本。这种平台可在水深 2134m（7000ft）的地方运行。

钻井船

钻井船是指装有钻井设备和动态定位系统的海上船舶，它可以将自身保持在油井上方。这类船主要用于勘探钻探，它可在深达 3700m（12000ft）的水下作业。

浮动式生产系统

FPSO（浮动式生产、储存和卸载系统）是主要的浮动式生产系统。可根据需求被用作浮动的半潜式平台或钻井船。该系统主要用于油气的加工和储存，适用于在深达 1829m（6000ft）的海水中作业。

对环境的影响

海上石油生产会产生环境风险，最突出的是在通过油轮或管道将石油从平台运输到岸上设施的过程中发生的石油泄漏，以及发生在平台上的泄漏和事故。此外，在采油过程中，随着石油和天然气一同被带到地面的废水也会影响环境，这类废水通常是高盐的，可能还包含溶解态的或难以分离的碳氢化合物。

4.2 波 浪 能

海洋能包括海浪、潮汐、盐度和海洋温差所携带的能量。全世界海洋中海水的运动蕴含着大量的动能。这些能量的一部分可以用来为全球的家庭、交通和工业发电。

波浪能又称海洋波浪能，是一种可再生能源，它利用波浪的能量发电。与潮汐能利用潮汐的涨落产生的流动不同，波浪能发电利用的是海水表面在波浪作用下的水平和垂直方向的周期性运动携带的动能。大部分的波浪能装置的运行原理是将海洋波浪的上下运动转化为可以被发电装置利用的机械能（这个发电装置通常被称为换能装置，或者简称为PTO），并以此驱动装置发电。图1展示了波浪能的全球分布情况。

那它具体是如何工作的呢？下面的示意图展示了波浪能的原理以及如何用它来发电。

基于上述波能转换原理，根据欧洲海洋能研究中心（EMEC）的定义，目前有几种典型的波浪能发电技术，最常见的描述如下：

衰减器

衰减器是一种沿波浪传播方向布置的浮式装置。这种装置的技术原理允许其从功率较低的入射波浪中提取波浪能。装置的浮式管段随着波浪的运动而运动，从而驱动用于换能发电的液压系统发电。所有的用于换能发电的机械设备均被置于水密管段的内部。这种装置通常由两个或更多的管段互相连接构成，并可以被部署在各种水深。互相连接的管段数量越多，发电量越大。

点吸收器

点吸收器也是浮式装置，但其构造允许从所有方向的入射波中获得能量。此外，该装置并不是很大，具有典型的点浮标的形状，从而允许几台同类装置构成阵列从同一个波中提取能量。浮标可以将入射波浪能转换为振子动能、液压机械能或气动能以驱动发电机。换能发电系统可以安装在浮标内，也可以布置到海底。为了保持装置的定位，点吸收器一般会锚固在海底，其电缆也从海底延伸上岸。点吸收器的发电效率一般在30%～45%。

振荡波浪涌转换器

振荡波浪涌波能转化装置利用波浪的纵荡运动发电。这类装置通常具有一个用于捕获波浪能的透水或不透水浮式拦截器，拦截器通过铰链固定在海底的基座上。波浪携带的动能被拦截器拦截，使得拦截器在水中绕其铰链转动，形成振荡。拦截器与基座间通常由液压、气动或者线性机械装置连接作为换能发电装置，从而将拦截器产生的动能转换为电能。考虑到这类装置的拦截器必须能够拦截海面附近的波浪能，这类装置一般只能部署在较浅的水中，否则拦截器将过大，降低装置的可靠性和效率。

振荡水柱

这个装置利用空气循环来驱动发电机。采用这类技术的波浪能装置既可以置于岸上（易于实现大装机容量），又可作为浮式装置部署于海上（相对而言装机容量较小）。该类装置通常具备一个封闭的空气气室，这个气室的下半部分与海相通，上半部分则保持气密性，只留一个连接换能装置的出口。波浪的作用会驱使气室内的水面上下运动，振荡的水面就像一个"无重量的活塞"，通过压缩和扩张气室容积制造气室内的压力振荡，从而驱使空气流经气室上部的唯一开口。开口附近的换能装置通常采用威尔斯涡轮或冲击式涡轮，使得无论空气是从气室流出，还是从外界吸入，都转换为涡轮传动轴上同一个方向的扭矩，因此降低了能量损失。

溢流式装置

溢流式波浪能装置通过一个斜坡捕获入射波爬高后产生的溢流，溢出的水体流入位置高于海面的水库内，从而将波浪携带的动能转化为重力势能。这些水随后在重力的作用下通过安装在水库底部的换能装置（通常是水轮机）向下流出并最终排回海中。为了提高这类装置的效率，可设计喇叭形的附属结构实现聚能的效果。溢流式装置可以被安装在岸线上，或者作为浮式装置安装在海里。一般而言，由于岸线附近存在波浪破碎导致的波浪能耗散，岸上的溢流装置的发电效率要低于部署于离岸的装置。

4.3 海 底 管 线

海底管线又称海水下、海洋或离岸管线，它通常铺设在海床或海床以下的沟槽中。有时管线主要在岸边，但在其他一些地方，它穿越了一些小水域，如海湾、海峡和河流。海底管线主要用于运输石油或天然气，水的输送也是一个同等重要的用途。

路线选择

海底管线规划中首先和最重要的任务是路线选择。这种选择必须考虑各种各样的问题，有些问题具有政治性，但其他问题主要涉及地质灾害、路线上的物理因素以及所涉及区域的海床的其他用途。

物理因素

海底管线建设中主要的物理因素是海床的状况，也就是说是平滑（即相对平坦）还是粗糙（存在不同高程点的起伏）。对于不均匀的海床，管道在连接两个高点时存在无支撑的自由跨度。如果无支撑管段太长，所承受的由于自重产生的弯曲应力就会变得过大。由水流导致的涡旋引起的振动也能造成问题。无支撑的管道跨度的矫正措施包括海床调平和安装后加固，例如在管道下方建护堤或填砂。海床强度也是一个重要的因素。如果土壤不够坚硬，管道可能一定程度上发生下陷，导致检查、维护和预设的连接实施起来变困难。在另一种极端情况下，岩石海床挖沟的成本很高，而且在高点可能会发生管道外部涂层的磨损并随后产生破坏。一种理想情况是，土壤可以使管线一定程度沉降于其中，从而为管线提供一些横向稳定性。

修建管道前应考虑的其他物理因素包括：

（1）海床移动性：沙波和大型沙纹这类海床特征会随着时间移动，因此在施工过程中由这些特征的顶部支撑的管线可能会在后期管道投入运营时被发现处于一个沟槽中。这些特征的发展很难估计，所以最好避免在它们存在的区域进行建设。

（2）海底滑坡：它们由发生在较陡峭的斜坡上的较高的沉积速率造成，也可以由地震触发。当管线周围的土壤发生滑动时，特别是当由此产生的位移与管线呈大角度时，土中的管道可能会遭受严重的弯曲和拉伸失效。

（3）水流：水流强度高是令人反感的，因为它阻碍了管线的铺设操作。例如，在浅海中，两个岛屿之间的潮流可能相当强。在这种情况下，即使新路线会更长也最好把管线铺设到其他地方。

（4）波浪：在浅水区中波浪（在极端波浪情况下）对于管道铺设作业是个很棘手的问题。由于波生流的掏刷作用，对建成后管线的稳定性也大为不利。这就强调了在管线登陆（管道到达海岸线的地方）时需要精心选址的重要性。

（5）海冰：在冰冷的海水中，浮冰经常会漂入浅水域，其下部会接触到海底。当继续漂移时，它们可能会掏蚀海床，继而撞到管道。冰柱也可以通过施加较大的局部应力或引起周围的土壤失效而破坏管道。冰眼是冷水中另一种对管道的危险因素，通过它喷出的海水可以移除管道下方的土壤，从而使管线容易受到由于自重造成的过大应力或涡流引起的振荡的影响。在确定存在风险的地区，可以在规划路线时将管道设计铺设在回填沟中。

管线施工

管线施工包括两个步骤：将大量的管段组装成一条完整的线路，随后沿所设计的路线安装该线路。可以使用几种方式铺设海底管线，方法的选择是基于以下原则：物理环境（例如：波浪、水流）、设备的可用性和成本、海水深度、管线长度和管道直径、与沿线已有的其他线路和结构相关的限制。现有的铺设方式一般分为四种：拉/拖式、S式、J式和卷筒式。

拉/拖系统

在该系统中，海底管道在岸上安装，然后被拖到指定位置。装配可以是平行或垂直于海岸线。对于前一种情况，整条管线可以在拖出和铺设前连接好。对于拖曳程序，可以使用下面一些方法：表面拖曳、近表面拖曳、中间水深拖曳和水底拖曳。

管线的稳定

有几种方法用来稳定或保护海底管线及其部件。这些方法可以单独或组合使用，如：挖槽、埋填、垫层、锚固等。挖槽可以在管道铺设（预铺前挖沟）之前进行，也可以通过从管道下方移土（铺设后挖沟）进行。对于后一种情况，挖槽设备安装在管道顶部或横跨管道。有一些系统可以被用来在海床上为海底管线开槽。例如射流装置是其中的一种系统，它采用铺设后的挖槽方式，通过使用强大的泵向管道的两侧喷水从而实现移除管线下方土方。

4.4 海洋灾害

海洋灾害是指发生在海洋中的自然灾害，是由正常海洋环境的急剧变化引起。海洋灾害包括从海岸线到海洋中的灾害。就像任何其他灾害一样，海洋灾害对人类是非常危险的，给社会、经济、财产和生命带来巨大的负面影响。下面介绍以下两种典型的海洋灾害：

风暴潮

风暴潮是指风暴期间海水的异常上升，其大小用相对天文潮位的高度来衡量。风暴潮主要是由于剧烈的大气扰动，如强风和气压骤变引起。风暴潮的大小取决于所在位置海岸线和风暴路径的关系、风暴的强度和大小、风暴移动速度以及海床地形。

热带气旋的压力影响导致海水面在低气压区域上升，而在高气压区域下降。水位的上升会抵消低气压，使水面下某个平面的总压力保持不变。据估计，大气压每下降1毫巴（百帕），海平面就会增加10mm。

除了上述过程之外，近岸地区的潮位和波高还受近底层水流的影响，即海床的形态和高程影响。

4.4 海洋灾害

影响风暴潮的因素

风暴潮是由沿着风暴中心旋转的风的作用推动向岸运动的水形成。相对于风往岸边吹的作用，由强风暴形成的低压对风暴潮位的贡献微乎其微。

特定地点的潜在的最大风暴潮取决于一些因素。风暴潮是一个非常复杂的物理过程，因为它对风暴强度、移动速度、大小（最大风速半径）、移动轨迹与海岸的夹角、中心气压（影响远小于风）以及海岸地貌特征都很紧密相关。

影响风暴潮的其他因素还包括大陆架的宽度和坡度，坡度较缓的大陆架相对陡峭的大陆架容易产生更大的风暴潮。例如，同样是一个四级风暴，如果在路易斯安那州海岸登陆，由于该海域大陆架宽而且浅，可能产生 20ft 高的风暴潮；而如果是在佛罗里达州的迈阿密海岸登陆，由于该地区大陆架坡度变化快，可能只会出现 8ft 或 9ft 的风暴潮。

风暴潮的破坏力会因为波浪而增加，波浪对海岸建筑物的连续冲击可能加剧对海岸建筑物的破坏。水的密度是 $1t/m^3$（1700 磅/立方码），海浪频繁且持续地作用在海岸结构物上，如果设计时未专门考虑这个力的作用，很可能导致结构物的毁坏。风暴潮和波浪的共同作用会增大对岸边的影响，因为水位增加导致波浪向岸传播范围增大。

此外，风和浪联合产生的水流作用会导致海滩和沿海公路遭到严重侵蚀。强风暴影响后的建筑物很可能因为地基遭受侵蚀而产生破坏。

海啸

海啸是由地震、海底火山爆发、海底滑坡、火山泥中气体的喷出、冰山倒塌、核爆炸，甚至是从太空坠落的物体引起的，由于水体或海床突然发生剧烈变化产生的一系列的长波。

海啸的速度取决于水深。在深海，海啸可以以 500~1000km/h 的速度移动。在近岸浅水中，速度降低到每小时只有几十千米。

海啸的高度也取决于水深。在深海，海啸的波幅通常仅有 1m 高，但在海岸线附近，波幅可高达几十米。

成因

海啸，通常错误地被认为是潮汐波，可由不同的方式激发。特定区域的海床抬升（地震）是海啸产生最常见的原因。这种海啸可以在消失前传播数千英里而且往往会造成大量的破坏。水下滑坡发生在大块沉积物沿海床移动时，是海啸产生的另一个常见原因。滑坡产生的海啸往往发生在相对局部，通常比地震产生的海啸造成的破坏小。最后，沿岸地区的火山和局部的滑坡也会引发海啸。然而，这些海啸发生在极有限的范围内，一般只影响一个核心地区。

海啸的"生存期"由三个阶段组成：产生、传播和爬高。其中我们最关注的是前两个阶段。在一个海啸预警系统中，我们努力做到最好地监测海啸的产生，并在海啸产生后了解其传播的方向。

海啸振幅很大程度上取决于两个因素：海床的地形和潮汐。海底地形通过改变海啸的波长和波高之间的比值来改变海啸的波高。一般来说，波长与波高的比值随着波浪进入较浅的水而降低，导致海啸尺度的增大。潮汐也会影响海啸对一个地区造成的破坏，高潮时海啸能够造成比低潮时大得多的破坏。

参 考 文 献

[1] 韩理安. 港口水工建筑物 [M]. 北京：人民交通出版社，2008.
[2] 胡旭跃. 航道整治 [M]. 北京：人民交通出版社，2017.
[3] 黄伦超，陶桂兰. 渠化工程学 [M]. 北京：人民交通出版社，2016.
[4] 季小梅，陈伟，申锦瑜. 港口航道与海岸工程专业英语 [M]. 北京：人民交通出版社，2019.
[5] 姚宇，杜睿超，袁万成. 港航专业英语教学面临的问题及其对策 [J]. 大学教育，2015（09）：101-102.
[6] 郑艳娜. 港口航道及海岸工程专业英语 [M]. 南京：东南大学出版社，2017.
[7] 邹志利. 海岸动力学 [M]. 4 版. 北京：人民交通出版社，2010.
[8] 朱梅心. 港口及航道工程专业英语 [M]. 北京：人民交通出版社，1995.
[9] Garrett N.，Kris S.，李道季. Advanced Geography through Diagrams 地理学专业英语基础（图示教程）[M]. 上海：上海外语教育出版社，2000.
[10] Coastal Engineering Manual [M]. U. S. Army Corps of Engineers，2013.
[11] Cruz J. Ocean Wave Energy：Current Status and Future Prespectives [M]. Springer Science & Business Media，2007.
[12] Dean R. G.，Dalrymple R. A. Coastal Processes with Engineering Applications [M]. Cambridge University Press，2004.
[13] Dean R. G.，Dalrymple R. A. Water Wave Mechanics for Engineers and Scientists [M]. World Scientific Publishing Co.，1992.
[14] Holthuijsen L. H. Waves in Oceanic and Coastal Waters [M]. Cambridge University Press，2007.
[15] Knighton D. Fluvial Forms and Processes a New Perspective [M]. Routledge，2014.
[16] Lambert M. S.，Mariam T. T.，Susan F. H.（Eds.）. Spar（Platform）[M]. Betascript Publishing，2011.
[17] Pierre Y. J. River Mechanics [M]. Cambridge University Press，2002.
[18] Thoresen C. A. Port Designer's Handbook [M]. ICE Publishing，2014.
[19] Thorpe T. W. A Brief Review of Wave Energy [M]. Harwell Laboratory, Energy Technology Support Unit，1999.
[20] Tsinker G. P. Handbook of Port and Harbor Engineering [M]. Springer，1997.